Developments in Cognitive Radio Networks

Bodhaswar TJ Maharaj
Babatunde Seun Awoyemi

Developments in Cognitive Radio Networks

Future Directions for Beyond 5G

 Springer

Bodhaswar TJ Maharaj
University of Pretoria
Pretoria, South Africa

Babatunde Seun Awoyemi
University of Pretoria
Pretoria, South Africa

ISBN 978-3-030-64655-4   ISBN 978-3-030-64653-0   (eBook)
https://doi.org/10.1007/978-3-030-64653-0

This Springer imprint is published by the registered company Springer Nature Switzerland AG
The registered company address is: Gewerbestrasse 11, 6330 Cham, Switzerland

To family

Foreword

The field of cognitive radio networks is an important field in the development and realization of modern and future wireless communication networks. As the technology evolves, it is necessary to have not just an up-to-date account of its current state in the evolution process but also a detailed insight into its prospects and directions, especially in its drive towards being a key player in the nearest future of wireless communications. Thankfully, this book titled, 'Developments in Cognitive Radio Networks: Future Direction for Beyond 5G' meets this need perfectly.

In the book, the authors provide a brief history on cognitive radio networks before delving into the important aspects of spectrum and resource realization and utilization for modern cognitive radio networks. The book exposes the most recent methods and models for resource optimization in cognitive radio networks and presents adequate analysis of these methods and models. It then examines the latest tools and techniques being employed to drive cognitive radio networks in the beyond 5G era, such as the concepts of queuing theory, cooperative diversity, stochastic geometry and deep learning. Ultimately, the book explores the promising prospects and applications of cognitive radio networks to most other emerging technologies such as fifth and sixth generation networks, internet-of-things, advanced wireless sensor networks, smart cities, fourth industrial revolution and many more. The book is thus comprehensive and compelling in its approach and answers and will definitely be worth the while for all open-minded readers, engineers and researchers keen on learning and advancing the field of cognitive radio networks.

The authors are well-known colleagues and seasoned researchers in the field of wireless communications. The authors' contributions to modern wireless communications are well covered in various IEEE conferences and journal article publications. Most of their works have been presented through patents, articles, book chapters and books in the well-established platforms in Electronic and Computer Engineering and Telecommunications.

It is therefore with deep pleasure that I introduce and welcome this book into the body of knowledge in the field of cognitive radio networks and trust that many will find this a useful text.

Johns Hopkins University, Baltimore, MD, USA   Ashutosh Dutta, Ph.D.

Fellow of the IEEE
IEEE Future Networks Founding Co-Chair
IEEE Communications Society Distinguished Lecturer
Member-At-Large—IEEE Communications Society
December 2020

Preface

For years now, the cognitive radio networks (CRN) has continued to evolve as a leading technology to help drive modern and near-future wireless communication possibilities. With its impressive prospects and intriguing promises, more efforts and means are being dedicated to studying and developing CRN models that can achieve outstanding results and remarkable performances. Notably, there are already a plethora of volumes on the CRN so much that the need for another book on this subject may be rightly questioned. In response, the authors' perspective is that, despite the sizeable number of volumes on the CRN, most volumes seem to have been too narrow or unintentional in their approach of the CRN. Indeed, there are works that have focussed on the sensing of unused or underutilised radio-frequency spectrum for possible CRN application. There are other works that have focussed on making the sensed spectrum available to help drive CRN operations, and so on. However, there is still a need for a concise but comprehensive book on the CRN that covers all the essential aspects, while also presenting the most recent research ideas and developments on the subject matter. This new book is designed to address that opportunity.

The first part of the book provides the necessary background on the CRN and extends to cover the important aspect of spectrum for the CRN. The spectrum is well established as the most important resource for the CRN. Because of the spectrum's importance in the CRN scheme, emphasis is laid on the need to discover and implement the most efficient techniques for sensing unused and/or underutilised spectrum for an effective CRN realisation. Then, important recent developments and new findings from various spectrum sensing efforts for CRN applications are discussed in depth. Also, the most recently advanced techniques for achieving optimal or near-optimal spectrum sensing for the CRN, particularly the aspects of *cooperative sensing* and *predictive sensing*, are generously explored.

The second part of the book presents the CRN as being *beyond just the spectrum*. While the spectrum is indeed very important for the CRN, there are several other resources—bandwidth, timeslots, data rates, transmission power and others—that must be equally considered for an effective CRN realisation. In that case, the spectrum must be jointly considered, alongside these other CRN resources, in order to achieve and provide optimal and near-optimal solutions to help realise the utmost for the CRN. This is covered under the broad aspect of resource allocation (RA) optimisation for the CRN. In that second part of the book, the most appropriate optimisation tools for RA in the CRN are exposed, new and/or improved RA models and solutions for modern CRN application are examined, and the performance analyses of the new RA models are extensively carried out.

In the final part of the book, the most recent developments in CRN modelling, applications, evaluations and eventual realisation, are explored. In this part, analytical concepts such as *queuing theory* and *stochastic geometry* and technical concepts such as *cooperative diversity* and *machine/deep leaning* are established as important new ideas and approaches that are being introduced into the modern RA models for the CRN. These relatively newly introduced concepts are specifically incorporated to help improve the design, analyses and solutions of the CRN models, to address the aspect of interference management and control, to mitigate the effects of interference and other limiting constraints, and/or to achieve an overall greater resource management and productivity for the CRN. In this part still, as part of the ongoing developments in the CRN, some of the significant areas in which the CRN is impacting and will continue to impact emerging technologies such as the fifth-generation and the internet-of-things, and its impact on the drive towards the realisation of smart cities and a globally interconnected world, are discussed. Important contributions on how to help fast-track these new technologies and possibilities through the CRN are graciously offered.

In all, the unique feature of this book is that it is able to concisely relate all the important aspects of the CRN—spectrum sensing, spectrum availability, resource optimisation and others—in one simple piece, thereby providing a more holistic perspective about the CRN than most other volumes that are available on the subject. The book stands out from others in that it distinctly integrates the *new and improved ideas on spectrum sensing* with the *recent, most viable solutions on resource optimisation*, as currently achievable for the CRN. The striking contribution of the book is therefore that it successfully *brings together under one title all the important tools needed to get the most from the CRN*. The authors envisage that this book will assist new researchers and graduate students in understanding the field of CRN better and also trigger new opportunities and future research directions in this fast evolving area of modern telecommunications.

University of Pretoria, Pretoria, South Africa   Bodhaswar TJ Maharaj

University of Pretoria, Pretoria, South Africa   Babatunde Seun Awoyemi
October 12, 2020

Acknowledgements

The authors are greatly indebted to the many research students that have studied in the Broadband Wireless Multimedia Communications (BWMC) research group at the University of Pretoria for their work and outputs that has assisted in compiling this book. The authors wish to thank Professor Attahiru S. Alfa for his invaluable suggestions, reviews and critique during the compilation of this book.

The funding and support from SENTECH SOC over the years is greatly appreciated for without which many students would not have had the opportunity to pursue graduate and postdoctoral studies within this research group.

The authors acknowledge the reviewers of our works in cognitive radio networks over the years as this enriched and challenged us to think ingeniously while compiling this book. The support and encouragement by our family, friends and colleagues that contributed in one way or the other will forever be cherished.

Contents

Acronyms

5G	Fifth generation
AI	Artificial intelligence
AMC	Adaptive modulation and coding
BnB	Branch-and-bound
BPP	Binomial point process
BPSK	Binary phase shift keying
BS	Base station
CR; CRs	Cognitive radio; Cognitive radios
CRN	Cognitive radio network
CSU	Cooperative secondary user
DL	Deep learning
DRL	Deep reinforcement learning
DSA	Dynamic spectrum access
FCC	Federal Communications Commission
HetNet	Heterogeneous networks
IoT	Internet of things
ILP	Integer linear programming
LMI	Linear matrix inequalities
LP	Linear programming
LTE	Long-term evolution
MARA	Margin adaptive resource allocation
MATLAB	Matrix laboratory
MDPs	Markov decision processes
MHCP	Matern hardcore point process
MIMO	Multiple input multiple output
MINLP	Mixed integer non-linear programming
ML	Machine learning
NLP	Non-linear programming
NOMA	Non-orthogonal multiple access
NP	Non-deterministic polynomial-time
NRT	Non-real time

OfCom	Office of communications
OFDM	Orthogonal frequency division multiplexing
OFDMA	Orthogonal frequency division multiple access
PCP	Poisson cluster process
PGFL	Probability generating functional
PHP	Poisson hole process
PPP	Poisson point process
PT; PR	Primary transmitter; Primary receiver
PU; PUs	Primary user; Primary users
QAM	Quadrature amplitude modulation
QoS	Quality of service
RA	Resource allocation
RT	Real time
RARA	Rate adaptive resource allocation
SG	Stochastic geometry
SGD	Stochastic gradient descent
SINR	Signal-to-interference plus noise ratio
SIR	Signal-to-inference ratio
SNR	Signal-to-noise ratio
SSU	Source secondary user
ST; SR	Secondary transmitter; Secondary receiver
SU; SUs	Secondary user; Secondary users
SUBS	Secondary user base station
WSN	Wireless sensor network
xG	Next-generation
YALMIP	Yet another LMI parser

Symbols

$K; k$	Total number of heterogeneous secondary users; k is used to identify a particular user
K_1	Number of category one secondary users; number of category two users is $K - K_1$ (or K_2)
$N; n$	Number of available OFDMA subchannels; n is used to identify a particular subchannel
L	Number of primary users
$H_{k,n}^c$	Channel gain between SUBS and SU at the kth SU over the nth subchannel
$H_{k,n}^s$	Channel gain from SSU to CSU at the kth SU over the nth subchannel
$H_{k,n}^r$	Channel gain from CSU to D at the kth SU over the nth subchannel
$P_{k,n}^s$	Transmit power from SSU to CSU at the kth SU over the nth subchannel
$P_{k,n}^r$	Power from CSU to D at the kth SU over the nth subchannel
$c_{k,n}$	Data rate at the kth SU over the nth subchannel
$c_{k,n}^s$	Data rate from SSU to CSU at the kth SU over the nth subchannel
$c_{k,n}^r$	Data rate from CSU to D at the kth SU over the nth subchannel
$c_{k,n,D}$	Data rate at the kth SU over the nth subchannel for direct transmission
$c_{k,n,C}$	Data rate at the kth SU over the nth subchannel when cooperation is employed
$P_{k,n,D}$	Transmit power at the kth SU over the nth subchannel for direct transmission
$P_{k,n,C}$	Transmit power at the kth SU over the nth subchannel when cooperation is employed
$P_{\max}$	Total transmit power at SUBS
$\mathbf{x}$	Bit allocation vector
$\mathbf{x}_I; \mathbf{x}_{II}$	Bit allocation vector for a category one SU; bit allocation vector for a category two SU
$\mathbf{b}$	Modulation order vector
$\mathbf{b}_I; \mathbf{b}_{II}$	Modulation order vector for category one SU; modulation order vector for a category two SU
$\mathbf{p}$	Power transmission vector

$\boldsymbol{p}_D$	Power transmission vector for direct communication
$\boldsymbol{p}_C$	Power transmission vector for cooperative communication
R_K	Minimum rate demand of a category one SU
γ_K	Proportional rate constraint for a category two SU

Part I
Fundamentals on Cognitive Radio Networks

Cognitive radio networks are highly rated among the most-promising emerging technologies for the near future. The basic definitions, descriptions and deliberations on cognitive radio networks are presented in this first part of the book. The spectrum is also discussed as the most essential pivot on which the cognitive radio networks revolves.

Chapter 1
Introduction to Cognitive Radio Networks

1.1 A Growing Demand for Wireless Communication

Wireless communication has become an integral part of our everyday life [1]. Having evolved over the years, modern wireless communication is gaining a wide acceptance and global appeal, most especially because of its tremendous benefits over other traditional methods of telecommunication. Some of the benefits of wireless communication over other traditional telecommunication methods can be summarised as: *ubiquity* (massive coverage), *mobility* (on-the-go telecommunication access), *capability* (can accomplish a lot more), *capacity* (accomplishes much more with less), *portability* (a continuing reduction in component/device sizes) and *affordability* (a sustained improvement in service costs). As a result, the global demand for modern wireless communication paradigms are on an explosive, exponential rise. As an example, it is estimated that, at end of year 2019, over 5 billion of the world population already have access to mobile wireless communication through the use of cellular phones [2].

The numerous benefits of modern wireless communication (coverage, capacities, capabilities, etc.) is driving a continued increase in global demand, even in the so-called developing countries and continents of the world [3]. The growing increase in global demand for modern wireless communication is necessitating the development of new and improved technologies that can meet such demand. It is safe to project that if the recent trends in modern wireless communication continue, very soon, there will most likely be an immense, almost insatiable 'outbreak' in modern wireless communication applications, operations and demands all over the world.

Therefore, in response to the growing global demand, modern wireless communication is continuing to develop and advance new technologies that have the wherewithal to meet such enormous demands. One of the relatively new wireless communication technologies that is being advanced to help achieve modern communication realities and to meet global telecommunication needs is the *cognitive radio networks* (CRN). With its rich history and towering prospects, the CRN is

© The Author(s), under exclusive license to Springer Nature Switzerland AG 2022
B. TJ Maharaj, B. S. Awoyemi, *Developments in Cognitive Radio Networks*,
https://doi.org/10.1007/978-3-030-64653-0_1

continuing to gain worthy interest and global attention as one of the most significant wireless communication technologies to help meet the world's telecommunication needs for the immediate and the near future.

1.2   History of Cognitive Radio Networks

Historically, the CRN can be seen to have evolved as an answer to an all-important question for modern wireless communication. The question centres around *how modern and evolving wireless communication technologies are to negotiate and realise the requisite spectrum resource to accomplish their promises and goals.* Quite frankly, the question on spectrum realisation for wireless applications is a very significant question for modern wireless communication. The answer to this question is a potential 'deal-maker' or 'deal-breaker' for most breakthrough wireless technologies for the immediate and the near future. Till date, the CRN is one of the most advanced technologies that provide clear and promising response on how to realise and engage the needed spectrum for driving new wireless communication possibilities for the immediate and near future.

As earlier mentioned, it is now crystal clear that the sporadically growing demand for modern wireless communication applications mandates that new technologies be developed to meet this continuously increasing demand. As new technologies emerge to meet the growing telecommunication need, there is *an expanding increase in demand for and use of the radio-frequency spectrum needed to drive such technologies.* The big challenge, however, is that the radio-frequency spectrum is generally a *limited, non-expanding and non-ubiquitous resource.* In fact, because of the never-ending demand and agitation from various interest groups, in most parts of the world, the spectrum is already a scarce resource. There is, as it currently stands, an ongoing crisis that is emanating due to the problem of *spectrum scarcity* for wireless communication applications.

The spectrum scarcity problem arises from the fact that the radio-frequency spectrum—a fixed and limited resource—is currently *adjudged to have already been overstretched in its allocation and usage,* especially in most of the technologically advanced parts of the world. Consequently, it would seem that the spectrum resource needed for meeting the demands and expectations of modern wireless communication is either unavailable or, at best, insufficient. This problem of spectrum scare and/or inadequacy is obviously one of the greatest threats to modern wireless communication being able to achieve their promises and possibilities.

In response, the present problem of spectrum scarcity has necessitated the review of the principles by which the limited spectrum resources are being assigned for use by the regulatory bodies saddled with the responsibility of allocating the spectrum resource. There are now a number of examples of such works on the review of spectrum allocation and usage designs and patterns, and several similar works are still being carried out in many parts of the world in this regard. Some of the reports on spectrum usage patterns as carried out by the Federal Communications

Commission (FCC) in the United States are readily available in references [4, 5]. There are also reports on spectrum usage from the Office of Communications (OfCom) in the United Kingdom [6, 7] and the Independent Communications Authority of South Africa (ICASA) [8], among others.

Interestingly, the various investigations into the spectrum allocation and usage patterns seem to have all come up with a similar finding. The underlying conclusion from the various investigations on the spectrum is that the limited spectrum resource has not only been *poorly allocated*, but that, in most cases, it has also been *inefficiently utilised* by the various networks to which it has been allocated for use. So, while there is indeed the problem of spectrum scarcity, an equally important problem that arose in the course of the numerous investigations and reports on the radio-frequency spectrum usage patterns is the existential *problem of spectrum underutilisation*.

Solving the problem of spectrum scarcity and underutilisation is very critical to achieving the promises of modern wireless communication. Efforts to address this problem are still being investigated. One important finding from the various reports on the spectrum is that the current strategy being employed for allocating the spectrum is grossly ineffective. Therefore, there is a great need to establish *better allocation strategies* for the scarce radio-frequency spectrum.

The current strategy being employed in allocating the spectrum is a *static allocation strategy*. With this strategy, the spectrum is divided into fractions or parts. Each fraction or part of the spectrum is allocated to a specific operator in a static manner, without any possibility of an overlap in allocation and usage. This static allocation strategy has the important advantage of reducing possible interference among the various operators. However, ongoing reviews on the spectrum allocation and usage revealed that the static allocation strategy is a highly ineffective one. This is because, most operators only use their allocated spectrum at certain times (maybe during the day or at pick periods), while at other times, the spectrum is mostly idle. Employing this static allocation strategy is actually what led to the current situation in which the already-limited spectrum is then being significantly underutilised.

With the static allocation strategy, despite the fact that the spectrum is being underutilised by most operators, spectrum regulators cannot reallocate an allocated spectrum to any other interested operator, due to strict regulations guiding such allocations. The regulators are constrained by the fact that since the operators have obtained licenses to use the spectrum, they must have unlimited access to the allocated spectrum all the time, whether or not the allotted spectrum is being fully engaged and/or in use. Besides, no other operator may be assigned such a spectrum, or have any kind of access whatsoever to such 'occupied' spectrum. This static allocation strategy has now been shown to be grossly inefficient because it creates multiple *spectrum holes* of unused and/or unoccupied spectrum spaces formed as a result of lack of activity on the allocated spectrum spaces at different times by must operators and occupiers of the spectrum.

In response to the problem described above, a new spectrum allocation strategy, called the *dynamic spectrum allocation/access* (DSA), is being advanced. The DSA proposes that the spectrum can be allocated in a more flexible or dynamic manner, such that a specific portion of the spectrum can be assigned to more than one operator/user at the same or at different times in similar or differing circumstances and for the same or different operations, provided that there are well-established guiding principles to govern such dynamics [9]. With the DSA, therefore, the spectrum is dynamically assigned and utilised. This makes the concept of double use, co-use and/or re-use of a spectrum space with multiple 'owners' meaningful and realistic [10].

Consequently, a number of modern wireless communication prototypes are now being designed and developed to employ the DSA in overcoming the spectrum scarcity challenge, and in achieving improved spectrum utilisation in their communication activities. *The CRN emerged as one of the most-important modern, newly evolving technologies to leverage the DSA*. Right now, the CRN is becoming an highly promising and very potent technology for the realisation of the new DSA paradigm. With the CRN, by leveraging the DSA, new and improved ways of sharing and utilising the spectrum are incorporated and employed in achieving greater resourcefulness and productivity for modern communication realisation.

Mitola is generally accredited for pioneering the works on the CRN [11, 12]. Because of his founding works on the CRN, Mitola has been referred to as the 'Father of CRN' [13]. Actually, the earliest available record of the use of the word 'cognitive radio (CR)' and the foremost clearest definition and description of CR are found in Mitola's work [11]. In his thesis in [12], Mitola described a new kind of radios which have cognitive capabilities. These new radios, initially referred to as *software defined radios* (SDR), have the capability to learn from their environment and to intelligently and dynamically adjust their operating parameters, based on what has been learned, to achieve better communication. In other words, SDR or CR enabled devices should be able to, among other things, dynamically adjust their frequency spectrum of operation to access/use new frequency spaces to suit their new environment or to meet their new demands. One thing that stands out clear from the ground old description of the CRN by Mitola is that the functionality of CRs and the CRN will greatly depend on the actual implementation of the DSA.

Since that pioneering work, more and more researchers and scholars in the field have continued to leverage the initial concepts and ideas from Mitola in further developing and describing CRs and the CRN, especially for practical, realistic implementations. One such well-known founding contributor whose works have been quite remarkable in the development of the CRN is Haykin [14]. Others like Doyle [15] and Akyildiz [16] have also had significant contributions to the development of the CRN. The important position that has been well established from the pioneering works on the CRN, and which is still being amplified through recent works on the CRN is that, in their ultimate design, CRs and the entire CRNs should be *capable of achieving much more than just dynamic access* and the ability to employ free or underutilised spectrum spaces for their communication [12, 14, 15]. The CRN must be beyond the spectrum. However, as the CRN evolves beyond the spectrum, this historical background on the CRN is important in that it shows in clear terms that it is the need to improve the use of the spectrum (being achieved by implementing a more flexible or dynamic mechanism for the allocation and usage of the spectrum resource) that formed the basic concept from which the CRN emanated.

1.3   Application of Dynamic Spectrum Access in Cognitive Radio Networks

In its design, the CRN leverages the DSA in achieving spectrum sharing and utilisation between a primary network and a secondary network. The primary network is made up of one or more original owners or primary users (PUs) of a spectrum space, while the secondary network is made up of some secondary users (SUs) of the same spectrum space [16]. The SUs are opportunistic users of the PUs' spectrum. A number of ways to design the primary-secondary networking for CRN have been described. A detailed description of various architectural designs for CRN is provided in the next chapter. In this section, however, a brief mention on possible designs for CRN is provided to help understand and establish the application of DSA in CRN.

In the foremost designs of DSA as applicable to the CRN, the SUs of the secondary network are to first *identify spectrum holes*. Spectrum holes are spectrum spaces that are free or available, despite being assigned to a primary network. The spectrum spaces are free because they are not being occupied or used by the PUs to which they have been assigned at those time instances. After identifying spectrum holes, the SUs must then *reconfigure themselves* to be able to use those frequencies of the spectrum hole to communicate or transmit their signal. Furthermore, as the SUs transmit their signal, they must *keep an eye on when the PUs return* and are ready to use their spectrum. The SUs must *immediately cease transmitting* in those spectrum frequencies and give room for the PUs to use them. Once the PUs re-occupy their spectrum spaces, the SUs using such spaces have to *identify new free spectrum spaces and reconfigure their parameters again* to use those new spaces. The SUs then continue their transmission in those newly available spectrum spaces. Again, these SUs have to be alert enough to identify the return of the PUs to those new spaces so they can vacate them and move to new ones. Interestingly, the entire process must happen as quickly and seamlessly as possible so that communication of the PUs and the SUs are not adversely affected.

Newer designs of DSA in the CRN permit the SUs to transmit their data and/or *communicate simultaneously with the PUs* at any given time. For this to happen, the primary-secondary network agreements have to have been reached beforehand. Such agreements usually mandate that the SUs communicate or send their data at low power over the entire bandwidth of the PUs (for example, in ultra wide band networks). This will ensure that the interference caused by the SUs to the PUs is at a bearable minimal. There could also be some *cooperative agreements* that may necessitate the SUs to help transmit some of the PUs' data, in exchange for a larger bandwidth, higher transmit power, longer transmission time or some other resource gain for the SUs. In fact, some more complex descriptions of the possible interplay between PUs and SUs in the CRN suggest that the SUs can *switch* between fully occupying the spectrum when the PUs are not there and going to the low transmission state when the PUs arrive. The various dynamics and possibilities that the CRN can achieve through DSA make the CRN an interesting technology. The CRN is therefore gaining the right recognition and attention as one of the most promising modern wireless communication technologies for the immediate and these near future.

1.4   Cognitive Radio Network Beyond Spectrum

Even though the CRN emerged as one of the technologies for achieving significant improvements in spectrum allocation and usage in modern wireless communication, further developments in its conceptualisation and application have established that the scope of the CRN must be beyond just the spectrum. In reality, modern communication technologies are generally being developed to have an enlarged scope so that they can have further appeal and farther reach. In that regard, the CRN is not an exception. The CRN must go beyond its interest in and ability to better manage or administer the spectrum.

As the CRN evolves, care is taken to ensure that the importance of the spectrum to the CRN is not undermined. However, the CRN must not be confined to the spectrum as it is, in fact, far and above spectrum sensing and/or spectrum availability alone. One of the clearest descriptions of this broader scope of the CRN, and how it transcends spectrum availability and usage, is seen in [15]. She described the scope of the CRN as being beyond just a technique to be employed for administering the spectrum in a more profitable manner. The author summarised the broader scope of the CRN as follows:

> [T]he CRN must be a self-organising system - it understands the context it finds itself in and can configure itself in response to a given set of requirements in an autonomous fashion. The configuration won't be on frequency or dynamic spectrum alone, but on other features too like power, beam pattern, routing algorithm, coding techniques, filtering techniques, etc. From the user point of view, the CRN will offer the benefit of personalising users' experiences so as to provide services tailored to the specific needs of individual users [15].

The enlarged scope of the CRN portends a wider, broader and more comprehensive usefulness for the CRN. With such expanded scope for the CRN, its usability will most likely become global. More so, the CRN will be applicable in almost all areas of human endeavour. The CRN will find applications in e-transportation, e-education, e-health, e-commerce, fourth, fifth and other industrial revolutions, automation, virtual reality, artificial intelligence, etc. Very clearly, therefore, the CRN will be an integral part of the modern society and/or smart cities and will be critical in the development and the eventual realisation of a highly interconnected world. The widened scope of the CRN also means that the technology will be a crucial element of global communication. This justifies the growing interest in its rapid development and almost eventual roll-out.

1.5   Possible Limitations with Cognitive Radio Network Applications

The CRN promises significant improvement in the capacities and capabilities of modern wireless communication. This is being achieved by providing ideas on how to engage or employ the spectrum resource to become more productive, among others. However, the CRN, as currently being developed, designed and projected, is not without some worrying signs and/or possible limitations. Some of the more specific limitations to the possible productivity of the CRN are identified in later chapters of this book that discusses certain concepts and principles of the CRN in-depth. However, to provide a brief but complete overview, some of the most generic limitations that are associated with CRN applications are briefly discussed in this section.

1.5.1   Resource Limitations

One of the greatest limitations to the effectiveness and productivity of the CRN is, maybe quite surprisingly, the limitation of the available resources needed to drive its operations [17, 18]. Since the CRN is designed to be predominantly an opportunistic network, the available resources to drive its operations are usually limited and, in many instances, non-guaranteed. This is a great threat to the CRN being able to achieve its ends. Quite frankly, unless the problem of resource limitations is addressed, the ideals of the CRN may never be fully realised. Addressing the problem of resource limitations in the CRN will imply that, especially for the secondary network, appropriate tools that can achieve effective, efficient and equitable distribution and utilisation of the limited CRN resources are established and incorporated in the network design.

1.5.2   Network Complexity

Another possible limitation to the CRN is the problem of network complexity [19]. The CRN is more complex than most other wireless communication networks because of the additional components, connections, processing and computational demands, etc. resulting from the primary-secondary networking for the CRN. Models that can minimise the effects of the extra complexities in CRN must be developed and implemented for the CRN. This will ensure that the CRN is a realistic and practicable wireless communication technology.

1.5.3   Problem of Interference

One other possible limitation for the CRN is the problem of interference [20, 21]. Interference has always been a great limitation for most wireless networks. Surely, this problem of interference is even more exacerbated in the CRN because of the primary-secondary network interplay that happens in the CRN. The PUs need to be guaranteed that the activities of the SUs will not cause undue interference and disrupt their network. The SUs too must be able to ensure that the activities of the PUs do not pose undue threat to their own communication. Solving interference problems in the CRN is very critical for its overall establishment and effectiveness.

1.5.4   Limitations of Wireless Communication

One important limitation for the CRN, as identified in this book, is the fact that the CRN, being a wireless communication design, like many others, must be built to cope with the common constraints and general limitations associated with all wireless communication networks. For instance, the limitation in the transmission power of the base station of a wireless communication network is a limitation that is a common denominator for all wireless communication networks, and for which the CRN must learn to cope with. For the CRN to achieve the utmost, therefore, such general limitations and constraints associated with wireless communication have to be factored in and carefully considered, and such limiting effects have to be effectively overcome.

1.6   Summary of the Chapter

In this chapter, we have established that the CRN is evolving as one of the most promising technologies for the near future, primarily because of the important promise of mitigating the challenge of spectrum scarcity/underutilisation. The CRN is leveraging the newly described DSA strategy in helping to address and mitigate the spectrum problem. As the CRN develops, the CRN will eventually achieve beyond just improving spectrum utilisation as its scope continues to expand. Even though it is a promising technology, the CRN is not without its own limitations. One of the most important limitations of the CRN, quite ironically, is the inadequacy of the resources available to drive the new technology. This is one of the most significant limitations of the CRN because it particularly relates to its resources and its resourcefulness. A significant part of this book is dedicated to seeking and establishing the most recent activities, attempts and approaches for addressing the resource limitations of the CRN. Several new interpretations, implementations and innovations for possible improvement in resource realisation and usage for the CRN are exposed and explored, making the book an important contribution to the body of knowledge in the field of CRN.

References

1. A. Goldsmith, *Wireless Communications*. Cambridge University Press, New York (2005)
2. Statista, Number of mobile phone users worldwide from 2013 to 2019 (2016). http://www. statista.com/statistics/274774/forecast-of-mobile-phone-users-worldwide/
3. P. Lange, 2012 Africa—mobile broadband, data and mobile media market (2012). http://www. budde.com.au/Research/2012-Africa-Mobile-Broadband-Data-and-Mobile-Media-Market. html
4. F.C. Commission, Report of spectrum efficiency working group. Spectrum Policy Task Force, Washington (2002)
5. F.C. Commission, Cognitive radio technologies proceeding, rep. ET Docket, no. 03-108, 2003
6. Ofcom, Securing long term benefits from scarce spectrum resources—a strategy for UHF bands IV and V (2012). http://wwwofcom.org.uk/consultations-and-statements/category-1/ uhf-strategy.
7. Ofcom, Techniques for increasing the capacity of wireless broadband networks—Ofcom (2012). http://wwwofcom.org.uk/static/uhf/real-wireless-report.pdf.
8. ICASA, Draft terrestrial broadcasting frequency plan 2013. Government Gazette, Republic of South Africa 574 (36321), 2013
9. C. Tran, R. Lu, A. Ramirez, C. Phillips, S. Thai, Dynamic spectrum access: architectures and implications, in *Proceedings of the IEEE MILCOM* (2008), pp. 1–7
10. J. Pastircak, J. Gazda, D. Kocur, A survey on the spectrum trading in dynamic spectrum access networks, in *Proceedings of the 56th International Symposium on ELMAR* (2014), pp. 1–4
11. J. Mitola, J. Maguire, G.Q., Cognitive radio: making software radios more personal. IEEE Pers. Commun. **6**(4), 13–18 (1999)
12. J. Mitola, Cognitive radio: An integrated agent architecture for software defined radios, Ph.D. dissertation, KTH, Sweden, 2000
13. A.M. Wyglinski, M. Nekovee, T. Hou, *Cognitive Radio Communications and Networks: Principles and Practice* Academic Press, London (2009)
14. S. Haykin, Cognitive radio: brain-empowered wireless communications. IEEE J. Sel. Areas Commun. **23**(2), 201–220 (2005)
15. L.E. Doyle, *Essentials of Cognitive Radio*, ser. The Cambridge Wireless Essentials Series. Cambridge University Press, New York (2009)
16. I.F. Akyildiz, W.-Y. Lee, M.C. Vuran, S. Mohanty, NeXt generation/dynamic spectrum access/cognitive radio wireless networks: a survey. Int. J. Comput. Telecommun. Netw. **50**(13), 2127–2159 (2006). http://www.sciencedirect.com/science/article/pii/S1389128606001009
17. B.S. Awoyemi, B.T.J. Maharaj, A.S. Alfa, Solving resource allocation problems in cognitive radio networks: a survey. EURASIP J. Wirel. Commun. Netw. **2016**(1), 176 (2016). https://doi. org/10.1186/s13638-016-0673-6
18. B. Awoyemi, B. Maharaj, A. Alfa, Optimal resource allocation solutions for heterogeneous cognitive radio networks. Digital Commun. Netw. **3**(2), 129–139 (2017). http://www. sciencedirect.com/science/article/pii/S2352864816301043

19. B.S. Awoyemi, B.T. Maharaj, Mitigating interference in the resource optimisation for heteroge-
    neous cognitive radio networks, in *Proceedings of the IEEE Second Wireless Africa Conference
    (WAC)* (2019), pp. 1–6
20. B.S. Awoyemi, B.T. Maharaj, A.S. Alfa, Resource allocation in heterogeneous cooperative
    cognitive radio networks. Int. J. Commun. Syst. **30**(11), e3247 (2017). https://onlinelibrary.
    wiley.com/doi/abs/10.1002/dac.3247
21. S.D. Okegbile, B.T. Maharaj, A.S. Alfa, Interference characterization in underlay cognitive
    networks with intra-network and inter-network dependence. IEEE Trans. Mobile Comput
    (2020). https://doi.org/10.1109/TMC.2020.2993408

Chapter 2
Perspectives on Cognitive Radio Networks

2.1 Architectural Descriptions of Cognitive Radio Networks

Architectural descriptions of the CRN provide fundamental ideas on how the CRN have been/are being designed to operate. The basic elements of the primary-secondary network design for a typical CRN are the primary users (PUs), the primary user base station (PUBS), secondary users (SUs) and, most likely, a control point or control unit for the SUs, usually referred to as the secondary user base station (SUBS) or access point [1]. In the design, PUs and the PUBS form the primary network while SUs and the SUBS form the secondary network. The communication and data transmission of the PUs is usually controlled by the PUBS, while the communication and data transmission of the SUs are controlled by the SUBS, whenever the SUBS are incorporated in the design. The various network architectural designs that have been described for the CRN are usually formed by the manner in which the elements of the CRN are aligned or combined. The diagram in Fig. 2.1 gives a general description of CRN architecture, indicating the manner in which the different components combine to form the basic CRN model.

2.1.1 Centralised, Distributed or Mesh

The most common architectural classification of the CRN considers the CRN as *centralised* (or infrastructure based), *distributed* (or ad-hoc based) or *mesh* (or combined) [2–5]. In all the three categories, the activities of the primary network are controlled by the PUBS. In the case of centralised architecture, *the SUBS, control unit or access point is responsible for coordinating and controlling the communication and data transmission of the SUs* in the secondary network. For distributed architecture, a SUBS or access point is not necessary. Hence, *the SUs simply communicate and transmit their data directly from one SU to another by*

© The Author(s), under exclusive license to Springer Nature Switzerland AG 2022
B. TJ Maharaj, B. S. Awoyemi, *Developments in Cognitive Radio Networks*,
https://doi.org/10.1007/978-3-030-64653-0_2

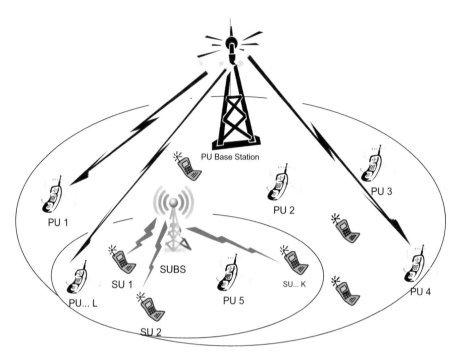

Fig. 2.1 A basic architectural description of the CRN

following defined guidelines. In this case, the SUs that are within a geographical space communicate with one another without the aid of a central communication hub. In the case of mesh architecture, the ideas of the centralised and distributed architectural designs are fused together to achieve the best of outcomes, usually at the expense of increased network complexity. Figures 2.2 and 2.3 give pictorial representations of the centralised and distributed architectural designs of the CRN. The mesh architecture is simply a fusion of these two designs and thus need no representation.

2.1.2 Overlay, Underlay or Hybrid

An equally important and widely used categorisation or classification of the CRN considers the CRN as *underlay* CRN, *overlay* CRN or *hybrid* CRN. This primary-secondary network architectural description of the CRN is influenced by the interference arrangements that exist between the primary network and the secondary network. This network architecture is a very crucial consideration in CRN modelling and practical implementation.

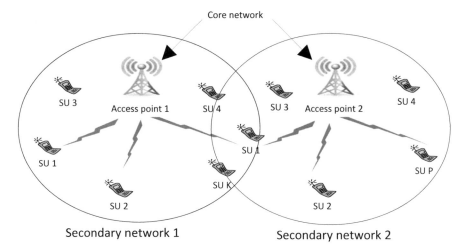

Fig. 2.2 A description of the centralised architecture for CRN

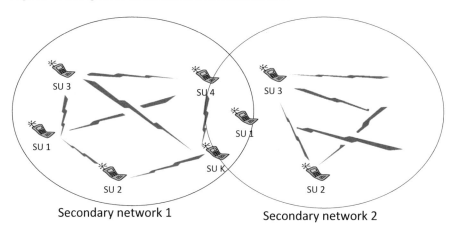

Fig. 2.3 A description of the distributed architecture for CRN

With underlay architecture, spectrum availability and usage is always guaranteed for all the PUs. At the same time, the spectrum space of the primary network is also to be used over the entire period of time by the SUs in the secondary network. The caveat, though, is that *the secondary network transmission may only interfere with the primary network transmission up to a pre-agreed interference limit* [6, 7]. The specific advantage that the underlay CRN architecture has over most other types of network designs is that it generates and provides substantial bandwidth for the secondary network, and that service provisioning for the secondary network may always be guaranteed all of the time (the possibility of service disruptions is quite low). The main disadvantage of underlay architecture is that the permissible

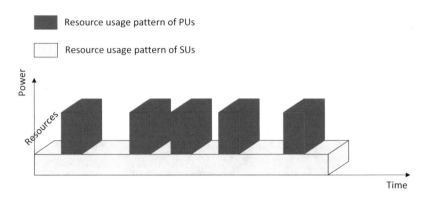

Fig. 2.4 A description of the resource usage pattern for the underlay architecture in CRN

interference limit of the primary network may be stringent and difficult to comply with. This tends to limit the productivity of the secondary network.

Figure 2.4 gives a simple description of the allocation and usage patterns of communication resources for the underlay CRN architecture. From the diagram, it can be observed that the SUs have access at all times to the resources (spectrum or frequency band, most especially), even at the periods when the PUs are transmitting or using the resources. However, the SUs are allowed to use the available resources to transmit at very low power, usually below the interference threshold of the PUs. This arrangement makes it possible for both the PUs and the SUs to co-exist and to co-use the resources *simultaneously*, as long as the interference is within the acceptable limit for the PUs.

In overlay architecture, the SUs cannot use the primary network's spectrum all the time. The SUs in the secondary network can only access the spectrum of the primary network when *the PUs are not available to use their spectrum or frequency bands* [8]. In those instances (of PUs' absence), the SUs have the liberty to use the primary network's spectrum maximally. As such, the SUs can communicate and transmit their data at high transmission power, data rates and/or modulation schemes. Once the PUs return, the SUs have to give way by immediately vacating the spectrum space so that the PUs can use those channels without interference.

The main advantage of overlay architecture is the high level of data transmission that the secondary network can achieve while the primary network's spectrum is available. However, a number of challenges have been identified with the overlay architecture. There is the problem of multiple service disruptions, whether at known or unknown intervals. Then, there is the issue of timing or sequencing in the ON-OFF interchange that happens when PUs arrive and SUs vacate a spectrum space, and vice versa.

The most common problems with the overlay CRN design are the problems of *miss detection* and *false alarm* [9]. A miss detection occurs when an SU determines, albeit wrongly, that a spectrum space is vacant or free, and that the PU occupying that channel is unavailable. Such an SU may then (be instructed to) go ahead to

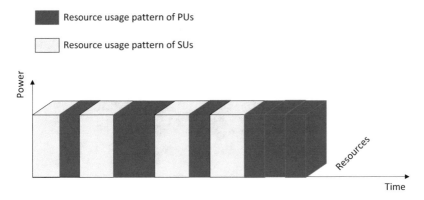

Fig. 2.5 A description of the resource usage pattern for the overlay architecture in CRN

transmit its data using the PU's spectrum. Since the SU (or secondary network control unit) is wrong in its assessment and the PU is actually using its spectrum, an unacceptable amount of interference is experienced by the PU. A false alarm occurs when an SU (or secondary network control unit) thinks that a PU is using its spectrum, when actually, that PU's channel is free. In this case, the SU does not (or is informed not to) transmit its data so as not to cause an undesirable amount of interference, while, in actual fact, the SU may have used that spectrum space unhindered since the PU is actually unavailable. The possibilities and effects of these misjudgements must be considered and well addressed and/or mitigated in practicable overlay CRN realisations.

Figure 2.5 gives a simple description of the allocation and usage patterns of communication resources for the overlay CRN architecture. From the diagram, it can be observed that the SUs have access to the resources (spectrum or frequency band, most especially) only at the periods when the PUs are not transmitting or using the resources. However, when the SUs have access to the resources, they are made to use the entire resources and can transmit at maximum power, without any fear of causing interference to the PUs. This arrangement means that SUs can only use the PUs' resources in an *opportunistic* manner, and they must be careful to not misjudge whether the PUs are present or absent, so that extreme interference is not experienced by the PUs.

In hybrid architecture, attempts are made to combine the benefits of the underlay and overlay architectural designs in one scheme, so that improved results are realised for the CRN [10]. In its design, when the PUs are unavailable and their spectrum vacant, the SUs communicate with the full transmission power available and at the highest rates achievable. Immediately the PUs are back, the SUs change to lower transmission power and data rates, as permissible by the PUs. In hybrid architecture, therefore, the SUs are made to benefit from the PUs' resources, whether the PUs are available to use them or not. As beneficial as the hybrid architecture for the CRN is, the major setback is the implication of a high network

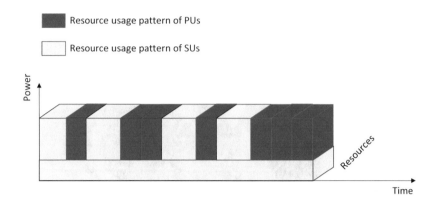

Fig. 2.6 A description of the resource usage pattern for the hybrid architecture in CRN

complexity. Simply put, the CRN as a hybrid network is much more complex to analyse and implement than the other two architectural depictions previously discussed.

Figure 2.6 gives a simple description of the allocation and usage patterns of communication resources for hybrid CRN architecture. From the diagram, it can be observed that the SUs have access to the entire resources (spectrum or frequency band, most especially) at the periods when the PUs are not available to use those resources, but when the PUs are available, the SUs use the resources in a very limited sense. Because of the flexibilities and continuous access to network resources that the hybrid CRN provides, hybrid architecture achieves the best results in resource usage for the CRN.

2.1.3 Cooperative or Non-cooperative

Another important architectural description of the CRN classifies the CRN as either cooperative CRN or non-cooperative CRN. One of the ways in which cooperative CRN architecture has been described is when there is some agreement among the SUs in the secondary network to work together or cooperate in their decision-making processes. This then implies that the SUs make multilateral decisions on aspects such as spectrum sensing, data transmission, resource management, etc. and all SUs take instruction from a central controller [11, 12]. In the non-cooperative CRN design, decisions on spectrum sensing, data transmission, resource management, etc. are unilateral and made by each SU.

In other cooperative CRN designs, cooperation happens between the primary network and the secondary network. In such cooperative CRN architecture, the PUs and the SUs decide to work together, and for one another. One way in which they may work together is that the SUs agree to first help transmit some of the PUs' data. Then, after the PUs' requests have been successfully carried out, the PUs allow their spectrum to be used by the SUs, or the SUs get some form of benefit from the PUs [13, 14]. In other words, if the SUs agree to support the PUs in driving the PUs' transmission or increasing the PUs' capacity, the SUs get the benefit of using a portion or the entirety of the PUs' assigned radio-frequency spectrum for the SUs to transmit their own data. Other important cooperative descriptions for the CRN are cooperative beamforming [15] and cooperative relaying [16]. With cooperative beamforming, the SUs cooperate to jointly use their antennas in making better and more accurate decisions on spectrum, data transmission, resource usage, etc. With cooperative relaying, the SUs cooperate to transmit each other's data, which helps to reduce the possibility and effects of high interference to the PUs.

The various architectural descriptions provided in this chapter are the commonest classes or categories of the CRN. A lot of research works on the various categorisations of the CRN so far provided are currently being undertaken. Furthermore, modern application models and designs of the CRN are being investigated, developed and deployed using these architectural descriptions and/or classification of the CRN.

2.2   Cognitive Radio Networks as Heterogeneous Systems

In the ongoing efforts towards developing the CRN, a recent but very crucial consideration is that the CRN must be developed as a heterogeneous and not a homogeneous system. Indeed, most of the earlier research works and experimental applications on the CRN have been carried out with the assumption that the network would be homogeneous. This is because the homogeneous CRN consideration is much easier to model, analyse, interpret and apply than the heterogeneous CRN consideration. However, it is almost certain that the CRN will be a heterogeneous system.

Recent works on the CRN now develop and study the CRN in the more practical and realistic consideration of it being heterogeneous. Especially in system modelling, to achieve a high level of accuracy in the description, interpretation and applications of the CRN, it must be considered as a heterogeneous system. By incorporating heterogeneity into the CRN, practicable CRN scenarios can be modelled, developed, analysed and eventually implemented. Investigating heterogeneity in the CRN is therefore a significant consideration, if the highest levels of network proficiency and productivity are to be realised for the CRN.

Several aspects of heterogeneity are being investigated for most modern wireless communication networks. For the CRN, aspects of heterogeneity that are most applicable are broadly classified under *heterogeneous networks*, *heterogeneous users* (or user demands) and *heterogeneous channels*.

2.2.1 Heterogeneous Network

In simple but clear terms, the CRN, by its design and application, is surely an example of a heterogeneous network. *Heterogeneous networks*, simply referred to as HetNet, is a relatively new research interest in the telecommunications research space. There are a number of ideas on the HetNet already and several others are on active investigation currently. However, the main idea on the HetNet, it seems, is that to build very robust and highly productive near-future wireless communication paradigms, there must be allowance for two or more networks to work together simultaneously over the same or on different communication resources and equipment, such as communication standards, base stations, radio access technologies, configuration parameters, architecture, transmission solutions, user demands, etc., in order to jointly expand their network capacity and reach [17].

A very good example of the HetNet consideration in the CRN is the primary-secondary networking in the CRN. A practical description of this primary-secondary network, as applicable to the CRN, is to consider a possible CRN scenario where a number of femtocells and/or picocells are made to work alongside a macrocell. In [18], a thorough study of the ideas and the technical issues that are familiar with HetNet has been carried out. As the CRN evolves, relevant elements and aspects of the HetNet must be incorporated in its design. Such relevant HetNet considerations, when incorporated into the CRN design, will influence the CRN in areas such as its resource problem formulation, and in attaining the desired level of accuracy in network realisation. Figure 2.7 provides a pictorial description of the application of HetNet in the CRN.

2.2.2 Heterogeneous Users or User Demands

The CRN must be developed to cater for a wide range of users demanding for resources and/or satisfying different kinds of applications and use cases. The idea of *heterogeneous users or user demands*, when applied to the CRN, means that the requirements or demands of one user may differ from another and/or other users. Therefore, the CRN, based on some defined principles, must be able to accommodate the need of each user or group of users without disparaging or disadvantaging other users [19, 20]. A number of things are to be considered when describing and classifying the various kinds of users and service demands for the CRN. The most important ones are briefly discussed.

2.2.2.1 Quality of Service Requirements

The SUs in the CRN may be classified using their minimum rate requirement that can guarantee an acceptable level of quality of service (QoS) for each user or user

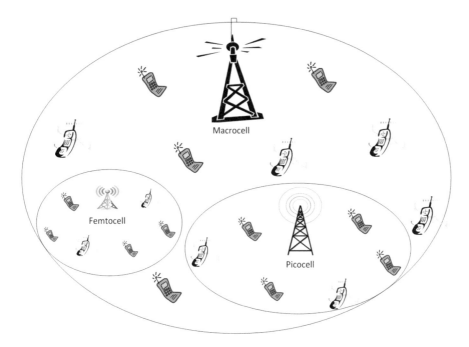

Fig. 2.7   A practical depiction of the HetNet for CRN application

category. If there are SUs that do not make any demands on an acceptable rate requirements, such SUs can be regarded as best effort service users. This kind of categorisation has been used for classifying the SUs in CRN designs developed in [7, 20, 21].

2.2.2.2   Service Type or Traffic Demands

The type of services being offered by SUs in the CRN may be used to categorise or classify them. In that case, services such as voice call, live-streaming, web surfing, background services like downloading, etc. may be employed in separating the SUs into relevant classes and meeting their needs accordingly. Some examples of the use of this type of categorisation of SUs are found in [22, 23].

2.2.2.3   Service Availability

The SUs in the CRN may be categorised based on whether or not they require services that are always on and that should never be interrupted. In that case, the SUs can be classified as either real-time (RT) or non-real-time (NRT) users. When the SUs in the CRN are classified as either RT or NRT, RT SUs are always given a higher priority in resource usage and service provisioning over the NRT users. Some examples of this method of classification are found in [24, 25].

2.2.2.4   User Sensitivity

The SUs in the CRN may be classified by considering how sensitive they are to certain stimuli. With this categorisation, for instance, if the sensitivity is based on their waiting time requirement, some users may be considered as being delay-sensitive (DS), while other users are said to be delay-tolerant (DT). There are several other ways of classifying users based on their sensitivity. Some examples of the use of this method of classifying SUs in the CRN are found in [23, 26].

2.2.3   Heterogeneous Channels

The CRN must be developed with a great deal of flexibility in its application and usage of the available channels and/or subchannels at its disposal. Therefore, *heterogeneous channels or subchannels* is an important consideration of heterogeneity in the CRN. In realistic CRN designs, the channels or subchannels that are available for the SUs will most likely not be around the same frequency range. The likelihood is that the frequencies that the SUs can employ would be located on separate parts of frequency bands, probably wide apart. Also, the different frequency channels may not all have the same properties. What it then means is that, at any given time, CRN devices must be able and ready to communicate using different channels or channel combinations to a heterogeneous set of devices within their vicinity [27].

In [28], the authors further explained channel heterogeneity in the CRN by stating that the channels in the CRN may not always be identical. The channels being different would imply that the propagation characteristics on each channel would differ. Furthermore, different channels would most likely support different types of data rates. Therefore, because the chances that the SUs in the CRN will have to use multiple channels or subchannels for their communication are quite high, the SUs in the CRN must be built to incorporate and use a heterogeneous set of radios for their communication and service propagation.

When designing, prototyping and implementing modern CRN systems, the aspect of channel heterogeneity must always be put into consideration. The channels and subchannels to be employed in the CRN must be designed to be in manageable sizes, with possible flexibility in their allocation and application, so as to achieve near-accurate representations of the CRN. By employing some of the recently advanced multi-carrier wireless transmission techniques, the realities of frequency hopping and mobility in the CRN may be well-taken care of. The orthogonal frequency division multiple access (OFDMA) and non-orthogonal multiple access (NOMA) techniques, and their more recent variants, are some useful multi-carrier techniques that can be employed for achieving channel heterogeneity in modern CRN designs.

The categories of heterogeneity thus far presented are the most common and the most applicable heterogeneous classifications for the CRN. Table 2.1 provides a concise summary of the various categories of heterogeneity, as applicable to the CRN.

2.3 Technologies to Drive Cognitive Radio Network

There are a number of other modern and/or newly developing wireless communication technologies and networks that are related to the CRN one way or another, and that would help in driving the CRN. One common denominator for all the newly emerging technologies (the CRN inclusive), which also relates them in a way is that *they seem to all have or share a common goal*. This common goal is that of making wireless communication the driving force for seamlessly interconnecting our developing modern cities and ultimately, the entire globe.

The drive to develop smart cities and to achieve a highly interconnected world can only be achieved by employing the right kind of technologies. These newly emerging technologies that are currently being developed, alongside the CRN, are grouped together under the umbrella called **next-generation (xG) wireless communication networks**. Some of the most prominent examples of xG networks are the *fifth-generation (5G) communication* and beyond, such as the *internet-of-things (IoT) networks* and the *next-generation wireless sensor networks (xWSN)*. A more detailed description of how the CRN relates to and will work well with these other emerging technologies to achieve smart cities and a highly interconnected world is provided in a later chapter of this book.

Importantly, the CRN and most of the other newly emerging technologies being developed will rely on and employ some improved technology-driven wireless communication techniques to help them achieve the level of networking required for optimal network productivity. Some of the new techniques that are currently being developed and implemented to drive and actualise the promises of the CRN and other newly-emerging technologies are discussed in the concluding parts of this chapter.

2.3.1 Cooperative Diversity and Relaying

Cooperative diversity and relaying are recent propositions for realising improved wireless channel conditioning for new technologies such as CRN and 5G. With the cooperative diversity and relaying techniques, cooperating users (also called relays or nodes), though geographically dispersed, use their antennas to form 'virtual' multiple input, multiple output (MIMO) systems. This makes it possible for the

Table 2.1 Concepts/classification of heterogeneity, as applicable to the CRN

S/N	Heterogeneous categorisation	Basis for classifying users	Basis for classifying users	Basis for classifying users	Basis for classifying users
1.	Heterogeneous networks	*Different standards*—GSM, EDGE, 3G, LTE, LTE-Advanced, etc.	*Different cell sizes*—macrocells, microcells, femtocells, picocells, etc.	*Cooperative networking possibilities*—direct communication, cooperative communication, relaying techniques, etc.	*Communication technologies*—wired, wireless, circuit-switched, packet-switched, etc.
2.	Heterogeneous users and/or user demands (or services)	*QoS or rate demands*—different minimum rates, different service rates, etc.	*Priority*—high priority (HP) users, low priority (LP) users, priority class (PC) users, best efforts (BE) users, etc.	*Sensitivity*—sensitive users, general users, etc.	*Delay profile*—delay sensitive (DS) users, delay insensitive (DI) users, delay tolerant (DT) users, delay intolerant (DI) users etc.
3.	Heterogeneous channels and/or subchannels	*Different channel bands*—channels and/or subchannels on different slices of frequency bands	*Different channel properties*—different channels and/or subchannels may have different properties	*Channel usage designs*—a single user should be able to use different channels and/or subchannels simultaneously	*Channel usage examples*—OFDM/OFDMA is a classic example of how heterogeneous channels can be applied in CRN

cooperating users to achieve diversity gains, even though the SUs are spatially dispersed [16].

When cooperative diversity or relaying is employed in modern wireless communication networks, cooperating users, though not in the same exact location, have some understanding and agreement to use their antennas to assist each other in the transmission (or retransmission) of their data to a particular destination user [29]. This is similar to conventional MIMO systems, therefore, cooperative diversity is sometimes referred to as virtual MIMO. Cooperative diversity and relaying have been shown to bring about a worthwhile increase in the achievable capability and network reliability of modern wireless communication networks, particularly the CRN and 5G networks [30].

2.3.2 Massive MIMO and Beamforming

Massive MIMO employs a high number of antennas, which are concentrated and used together within a small space in a transmitter to transmit a given signal. At the receiver, a large number of antennas are also used to receive multiple versions of the transmitted signal. The multiple received signals are then combined using some combination techniques to arrive at the best version of the transmitted signal and to minimise the need for signal retransmission due to a poor reception.

When MIMO techniques are combined with improved beamforming techniques, the high number of antennas is well focussed to help concentrate energy into a very small space. The result is that both the network throughput and the efficiency of radiated energy are significantly improved [31]. The development and implementation of modern wireless communication technologies such as the CRN, 5G and IoT networks will depend greatly on massive MIMO and beamforming techniques. Some of the benefits that these new technologies will derive from massive MIMO and beamforming are network robustness, efficient use of the spectrum, energy efficiency and improved network security [32, 33].

2.3.3 Cloud Computing

Cloud computing is one important technique to help drive modern wireless communication, especially the CRN, IoT and 5G networks. The large capacities and capabilities, fast processing speeds, etc. that the CRN and other new technologies promise will most likely result in a massive amount of data being generated. The low latency expectations of modern communication networks mean that the large data that are generated have to be transmitted as quickly as possible.

With cloud computing, provision is made for a large portion of the generated data to be stored in the cloud. Furthermore, significant portions of the data processing can be done in the cloud, which reduces the computational demand on the user equipment [34]. This means that newly emerging technologies, such as the CRN and autonomous vehicle-to-vehicle networks, will depend on cloud computing to help store and process significant portions of their data [35, 36].

2.3.4 *Orthogonal and Non-orthogonal Multiple Access*

Current wireless technologies, such as LTE and LTE-Advanced, have successfully employed orthogonal frequency division multiplexing (OFDM) and OFDMA techniques, and their variants, in driving their communication. These multiple access techniques have been shown to have promising prospects for the CRN as well. However, as promising as these techniques are, their applicability as the 'best' multiplexing and multiple access techniques for the CRN and many other emerging technologies is still in doubt.

The main challenge with OFDM/OFDMA and its variant techniques is that they usually have the challenge of singular channel allocation. What this means is that, each channel or subchannel can only be assigned to, and employed by, one user at a time [37]. To address this limitation of singular channel allocation, a very recent technique called the non-orthogonal multiple access (NOMA) technique has been proposed. The NOMA technique is being shown to be a better access technique for modern wireless technologies such as the CRN. This is because the NOMA technique attempts to overcome the singular channel allocation challenge associated with OFDMA techniques. To overcome the problem of singular channel allocation, the NOMA technique allows multiple users to communicate concurrently over a given resource by simply applying or following some non-orthogonal sharing principles [38, 39].

2.4 Summary of the Chapter

To summarise this chapter, we have shown that the modern CRN must be dynamic enough to blend into various architectural designs and must be capable of meeting various heterogeneous needs or demands. Also, the chapter discusses some of the newly developing technologies and new wireless communication techniques that will all play significant part in driving the CRN and in helping it actualise its many objectives and promises.

References

1. K.-C. Chen, Y.-J. Peng, N. Prasad, Y.-C. Liang, S. Sun, Cognitive radio network architecture: part I—general structure, in *Proceedings of the Second International Conference on UIMC*. ACM, New York (2008), pp. 114–119. http://doi.acm.org/10.1145/1352793.1352817
2. K.-C. Chen, Y.-J. Peng, N. Prasad, Y.-C. Liang, S. Sun, Cognitive radio network architecture: part II—trusted network layer structure, in *Proceedings of the Second International Conference on UIMC*. ACM, New York (2008), pp. 120–124. http://doi.acm.org/10.1145/1352793.1352818
3. C. Xin, X. Cao, A cognitive radio network architecture without control channel, in *Proceedings of the IEEE GLOBECOM* (2009), pp. 1–6
4. D. Xu, Q. Zhang, Y. Liu, Y. Xu, P. Zhang, An architecture for cognitive radio networks with cognition, self-organization and reconfiguration capabilities, in *Proceedings of the IEEE VTC (Fall)* (2012), pp. 1–5
5. A. Amanna, J. Reed, Survey of cognitive radio architectures, in *Proceedings of the IEEE SoutheastCon* (2010), pp. 292–297
6. M. Monemi, M. Rasti, E. Hossain, Characterizing feasible interference region for underlay cognitive radio networks, in *Proceedings of the IEEE ICC* (2015), pp. 7603–7608
7. B.S. Awoyemi, B.T. Maharaj, A.S. Alfa, Resource allocation for heterogeneous cognitive radio networks, in *Proceedings of the IEEE WCNC* (2015), pp. 1759–1763
8. W. Guo, X. Huang, Maximizing throughput for overlaid cognitive radio networks, in *Proceedings of the IEEE MILCOM* (2009), pp. 1–7
9. W.-L. Chin, J.-M. Lee, Spectrum sensing scheme for overlay cognitive radio networks. IET Electron. Lett. **51**(19), 1552–1554 (2015)
10. S. Senthuran, A. Anpalagan, O. Das, Throughput analysis of opportunistic access strategies in hybrid underlay overlay cognitive radio networks. IEEE Trans. Wirel. Commun. **11**(6), 2024–2035 (2012)
11. J. Lai, E. Dutkiewicz, R.P. Liu, R. Vesilo, Comparison of cooperative spectrum sensing strategies in distributed cognitive radio networks, in *Proceedings of the IEEE GLOBECOM* (2012), pp. 1513–1518
12. M. Nabil, W. El-Sayed, M. Elnainay, A cooperative spectrum sensing scheme based on task assignment algorithm for cognitive radio networks, in *Proceedings of the International Conference on WCMC* (2014), pp. 151–156
13. H. Xu, B. Li, Efficient resource allocation with flexible channel cooperation in OFDMA cognitive radio networks, in *Proceedings of the IEEE INFOCOM* (2010), pp. 1–9
14. H. Xu, B. Li, Resource allocation with flexible channel cooperation in cognitive radio networks. IEEE Trans. Mobile Comput. **12**(5), 957–970 (2013)
15. M.H. Hassan, M. Hossain, Cooperative beamforming for cognitive radio systems with asynchronous interference to primary user. IEEE Trans. Wirel. Commun. **12**(11), 5468–5479 (2013)
16. B.S. Awoyemi, B.T. Maharaj, A.S. Alfa, Resource allocation in heterogeneous cooperative cognitive radio networks. Int. J. Commun. Syst. **30**(11), e3247 (2017). https://onlinelibrary.wiley.com/doi/abs/10.1002/dac.3247
17. S. Landstrom, A. Furuskar, K. Johansson, L. Falconetti, F. Kronestedt, Heterogeneous networks (HetNets)—an approach to increasing cellular capacity and coverage, in *Proceedings of the 15th International Symposium on WPMC* (2012), pp. 108–112
18. D. Lopez-Perez, I. Guvenc, G. de la Roche, M. Kountouris, T.Q.S. Quek, J. Zhang, Enhanced intercell interference coordination challenges in heterogeneous networks. IEEE Trans. Wirel. Commun. **18**(3), 22–30 (2011)
19. S. Wang, Z.-H. Zhou, M. Ge, C. Wang, Resource allocation for heterogeneous cognitive radio networks with imperfect spectrum sensing. IEEE J. Sel. Areas Commun. **31**(3), 464–475 (2013)

20. R. Xie, F. Yu, H. Ji, Dynamic resource allocation for heterogeneous services in cognitive radio networks with imperfect channel sensing. IEEE Trans. Veh. Technol. **61**(2), 770–780 (2012)
21. B.S. Awoyemi, B.T. Maharaj, A.S. Alfa, QoS provisioning in heterogeneous cognitive radio networks through dynamic resource allocation, in *Proceedings of the IEEE AFRICON* (2015), pp. 1–6
22. M. Kaplan, F. Buzluca, A dynamic spectrum decision scheme for heterogeneous cognitive radio networks, in *Proceedings of the 24th International Symposium on ISCIS* (2009), pp. 697–702
23. B. Awoyemi, B. Maharaj, A. Alfa, Optimal resource allocation solutions for heterogeneous cognitive radio networks. Digital Commun. Netw. **3**(2), 129–139 (2017). http://www.sciencedirect.com/science/article/pii/S2352864816301043
24. S. Wang, M. Ge, C. Wang, Efficient resource allocation for cognitive radio networks with cooperative relays. IEEE J. Sel. Areas Commun. **31**(11), 2432–2441 (2013)
25. A. Alshamrani, X. Shen, L.-L. Xie, QoS provisioning for heterogeneous services in cooperative cognitive radio networks. IEEE J. Sel. Areas Commun. **29**(4), 819–830 (2011)
26. C. Shi, Y. Wang, P. Zhang, Joint spectrum sensing and resource allocation for multi-band cognitive radio systems with heterogeneous services, in *Proceedings of the IEEE GLOBECOM* (2012), pp. 1180–1185
27. M. Ma, D.H.K. Tsang, Impact of channel heterogeneity on spectrum sharing in cognitive radio networks, in *Proceedings of the IEEE ICC* (2008), pp. 2377–2382
28. V. Bhandari, N.H. Vaidya, Heterogeneous multi-channel wireless networks: routing and link layer protocols. SIGMOBILE Mobile Comput. Commun. Rev. **12**(1), 43–45 (2008). http://doi.acm.org/10.1145/1374512.1374526
29. B. Awoyemi, T. Walingo, F. Takawira, Predictive relay-selection cooperative diversity in land mobile satellite systems. Int. J. Satellite Commun. Netw. **34**(2), 277–294 (2016). https://doi.org/10.1002/sat.1118
30. A. Afana, E. Erdogan, S. Ikki, Quadrature spatial modulation for cooperative MIMO 5G wireless networks, in *2016 IEEE Globecom Workshops (GC Wkshps)* (2016), pp. 1–5
31. E.G. Larsson, O. Edfors, F. Tufvesson, T.L. Marzetta, Massive MIMO for next generation wireless systems. IEEE Commun. Mag. **52**(2), 186–195 (2014)
32. G. Liu, X. Hou, J. Jin, F. Wang, Q. Wang, Y. Hao, Y. Huang, X. Wang, X. Xiao, A. Deng, 3D-MIMO with massive antennas paves the way to 5G enhanced mobile broadband: from system design to field trials. IEEE J. Sel. Areas Commun. **35**(6), 1222–1233 (2017)
33. F.W. Vook, A. Ghosh, T.A. Thomas, MIMO and beamforming solutions for 5G technology, in *2014 IEEE MTT-S International Microwave Symposium (IMS2014)* (2014), pp. 1–4
34. D. Wubben, P. Rost, J.S. Bartelt, M. Lalam, V. Savin, M. Gorgoglione, A. Dekorsy, G. Fettweis, Benefits and impact of cloud computing on 5G signal processing: flexible centralization through cloud-RAN. IEEE Signal Process. Mag. **31**(6), 35–44 (2014)
35. A. Falchetti, C. Azurdia-Meza, S. Cespedes, Vehicular cloud computing in the dawn of 5G, in *2015 CHILEAN Conference on Electrical, Electronics Engineering, Information and Communication Technologies (CHILECON)* (2015), pp. 301–305
36. M. Tao, K. Ota, M. Dong, Foud: integrating fog and cloud for 5G-enabled V2G networks. IEEE Netw. **31**(2), 8–13 (2017)
37. L. Lei, D. Yuan, C. K. Ho, S. Sun, Power and channel allocation for non-orthogonal multiple access in 5G systems: tractability and computation. IEEE Trans. Wirel. Commun. **15**(12), 8580–8594 (2016)
38. Z. Ding, M. Peng, H.V. Poor, Cooperative non-orthogonal multiple access in 5G systems. IEEE Commun. Lett. **19**(8), 1462–1465 (2015)
39. X. Liu, Y. Liu, X. Wang, H. Lin, Highly efficient 3d resource allocation techniques in 5G for NOMA enabled massive MIMO and relaying systems. IEEE J. Sel. Areas Commun. **35**(12), 2785–2797 (2017)

Chapter 3
Spectrum Resource for Cognitive Radio Networks

3.1 The Place of Spectrum in the Overall Cognitive Radio Network Scheme

The CRN, alongside other modern and near-future wireless communication prototypes, will be instrumental in driving global interconnectivity, especially in this new era of smart cities and an interconnected world. Spectrum will be needed to make the many promises and possibilities of modern wireless communication materialise. The limitation in spectrum availability to accommodate the rising communication prospects is one of the most challenging problems that modern wireless communication face [1].

Without doubt, the spectrum holds a very special place and plays a highly significant role in the overall CRN scheme. The spectrum is a fundamental part of the CRN. Simply put, the CRN cannot exist or operate without the spectrum. There are two aspects of the spectrum that greatly affects the operation of the CRN, which are, *spectrum sensing* and *spectrum utilisation*. Spectrum sensing helps to identify the spectrum opportunities that are available for the CRN. Spectrum utilisation advances how the spectrum opportunities that have been identified are to profitably engaged for the CRN.

The performance of the CRN is significantly affected by these two aspects of the spectrum. Both spectrum sensing and spectrum utilisation influence the choice of the parameters for operation in the CRN such as the modulation schemes, transmission power, data rates and forward error correction coding rates. Actually, if the spectrum sensing and spectrum utilisation processes are inaccurately identified and executed, they pose very detrimental consequences to the operations of the CRN. The consequences of inaccurate spectrum sensing and poor spectrum utilisation in the CRN would include inefficient channel selections, preventable delays, suboptimal throughput realisation and an overall poor performance for the CRN.

More so, it is usually the information obtained during the spectrum sensing processes that is used for other aspects of the CRN such as channel selection, chan-

B. TJ Maharaj, B. S. Awoyemi, *Developments in Cognitive Radio Networks*,
https://doi.org/10.1007/978-3-030-64653-0_3

nel prediction and resource distribution and management (the aspect of resource allocation for the CRN is discussed in a latter chapter of the book). Therefore, if the spectrum sensing and spectrum utilisation processes are faulty, several other aspects of the CRN will be negatively affected. Because of its importance, therefore, the spectrum must be given a high priority in the overall CRN scheme.

3.2 Historical Context on the Spectrum in Cognitive Radio Networks

From an historical perspective, the spectrum has always been the most critical part of the CRN. The foundational concepts and ideas on the CRN are generally easily traced to the spectrum. It is greatly impossible to properly study, analyse and/or implement the CRN without adequate reference to the spectrum. It is important to reconsider how the CRN links to the spectrum, even as far back as the earlier works and thoughts on the CRN's development and evolution.

We already established in the first chapter of this book that the problem of spectrum scarcity and underutilisation—which arose because of the inefficient allocation and usage of the limited spectrum resource—is what birthed the CRN in the first place. The CRN became one of the earliest and most promising technologies developed to address this problem of spectrum scarcity/underutilisation. The CRN achieves this promise by implementing the DSA as a better and more resourceful alternative to the poorly implemented static spectrum allocation design that originally brought about the problem of poor utilisation and eventual scarcity of the spectrum [2, 3].

The CRN, through the DSA, provides the platform for multiple users to jointly access and use the same spectrum in communicating and transmitting their data. This brings about an improvement in the overall spectrum allocation and utilisation for the CRN. In essence, the CRN simply employs the possibilities and principles of the DSA to establish new and improved application models that optimise the use of the spectrum.

The foremost/foundational work on the CRN was carried out by Mitola [4]. Mitola's original description of the content and possibilities of the CRN clearly provides a sound historical perspective on the inseparable link between the spectrum and the CRN [5]. More so, the description by Doyle in [6] further supports this context. From contextual descriptions of the CRN in those works, what is clear is that the ideal cognitive device or user in a typical CRN system must be able to, among other things, change its frequency spectrum of operation in a dynamic manner to access/use the spectrum spaces that are available and that are best suited for different environments to achieve their desired communication goals.

For the above-mentioned attributes to be possible and/or achievable for the CRN, the DSA has to be implemented for the CRN. Although, it must be said that, in the eventual CRN design, cognitive devices or users in the CRN would not be defined only by their ability to access and use different spectrum spaces in carrying out their communication or data transmission. Definitely, the cognitive devices or users in the CRN will be much more than just spectrum hoppers. The simple point that is being emphasised is that the CRN, by employing the newly adopted DSA, will significantly improve the allocation and usage of the spectrum for its operation.

The inseparable link between the spectrum and the CRN shows that one greatly influences the other. The nexus can be described this way: *the CRN will make better use of the previously underutilised spectrum; meanwhile, spectrum exploration, availability and usage will be greatly improved by the CRN.* The CRN's first mandate is therefore to be able to make an adequate amount of spectrum available (by discovering sufficient spectrum holes) for practical CRN operations and applications. Fulfilling this mandate is greatly dependent on the ability to rightly *detect* or *sense* the spectrum holes in the network. The spectrum (through the DSA) and the CRN are thus fundamentally intercepted and significantly intertwined.

3.3   Spectrum Sensing in Cognitive Radio Networks

Spectrum availability and usage for the CRN is greatly dependent on accurate and adequate spectrum sensing. In many CRN designs, the secondary users (SUs) must be able to successfully detect both the activities of the primary users (PUs) as well as the presence of spectrum holes, before they may operate or carry out their activities. This is achieved through *spectrum sensing*. Generally, SUs in the CRN use information that they gather during the spectrum sensing process to make decisions about their activities. This makes spectrum sensing a very crucial component or aspect of the CRN. It is through spectrum sensing that the understanding of the radio environment that is required for an effective CRN operation is discovered, which makes it possible for its communication to be effectively carried out.

To achieve the most accurate and practicable spectrum sensing results for the CRN, some spectrum sensing techniques have been promulgated and are being implemented and applied in most CRN designs. Each technique has its own peculiarities and challenges, as well as merits and demerits. Some of the well-known techniques are briefly discussed.

3.3.1   Energy Detection Techniques

One of the most advanced and frequently used approaches for carrying out spectrum sensing in the CRN is the energy detection technique. This technique has been shown to have the lowest complexity of the approaches for spectrum sensing. However, the approach has also been shown to be the least accurate of the spectrum sensing approaches. With energy detection, a threshold value of signal energy value that indicates the presence of a PU is first determined. Then, an energy detector is used to measure the signal strength of the PU channels at intervals. The measured signals are compared to the threshold value and the results are used to determine the presence or absence of a PU on those channels. The advantage of the energy detection technique is that it does not need to have prior information about the PU signal. Simply, a band pass filter is used to pre-filter the measured signal. The signal is then squared and integrated over an interval of time so as to measure the amount of energy contained within the received waveform [7].

The biggest challenge with the energy detection techniques is that they have been argued to not be as accurate as most other spectrum sensing approaches. One other problem with energy detection techniques is that they do not perform well in low signal-to-noise ratio (SNR) conditions and under Rayleigh fading channel conditions [8]. Then, when energy detection techniques are used for detecting spread spectrum signals, they do not give a good sensing efficiency. However, their benefits of minimal computational demand and comparative ease of implementation still make them to be fairly accepted and widely used for spectrum sensing in the CRN.

3.3.2 Matched Filter Detection Techniques

The matched filter detection technique is another important detection technique for the CRN. In this technique, the secondary network devices in the CRN usually have the knowledge of some of the PU's signalling features beforehand. They then use this knowledge to check if a PU is present or not, and can even choose to demodulate the actual PU signal. The matched filter detection techniques represent the most optimum schemes for detecting PUs especially if the properties of the PUs' signal are well known by the secondary network devices [9]. In essence, matched filter detection is likely to be the most accurate technique for detecting the presence of a PU in the CRN scenario. The challenge with the matched filter techniques, however, is that they usually suffer from impractical levels of receiver complexity [10].

With matched filter detection techniques, the level of complexity required at the secondary network is usually difficult to meet in practical CRN designs. This is because, the foreknowledge of the signalling features of the PUs that the secondary network devices will be required to have will be quite enormous. More so, the secondary network devices must be capable of demodulating a number of different signal types. Thus, large samples of PU signals will be required to determine whether a specific band is occupied or free. The large number of samples required becomes an inherent limiting factor and, as a result, small periods of band usage cannot be sufficiently analysed. This is where the issue of the unrealistically high demands of receiver complexity arise from. This also poses a practical limitation on the time it will take to set up a network connection, which is a serious challenge for the CRN.

3.3.3   Cyclo-stationary Feature Detection Techniques

The cyclo-stationary feature detection technique is another technique that is well used for spectrum sensing in the CRN. This technique employs certain features of the PU signal that varies in a cyclic manner to determine whether a PU is present or not. Furthermore, some statistical attributes of PU signals, especially the mean, variance and autocorrelation of the signal, are used for detection purposes. Then, the presence of noise is distinguished from the actual modulated signals by using the fact that noise is stationary in a wide sense. This fact can also be used to separate one PU signal from another. The distinction in signals can be achieved by studying the cyclic spectral density function of the PU signals. Thus, the cyclo-stationary feature of each signal is used to simply distinguish between noise and a PU signal, and/or between two or more different PU signals [11].

The advantage that cyclo-stationary feature detection techniques have over other techniques for spectrum sensing in the CRN is that they are quite useful in situations when the amount of noise in the system is not pre-known. Therefore, the cyclo-stationary detection techniques have the advantage of working well even when noise and channel attenuation in the network are somewhat high. The challenge with these techniques, however, is that the sampling rate requirement is usually high. This may result in huge computational demand, which may make them impracticable in some instances and scenarios of the CRN.

3.3.4   Waveform-Based Sensing Techniques

One other technique of spectrum sensing that is well used in the CRN is the waveform-based sensing technique. This technique is best employed if the SUs are fully aware of the waveform patterns of the PUs. The SUs can then correlate the waveform patterns of the PUs with their own measured signals to determine whether or not a PU is available [12]. A number of non-statistical patterns may be used to carry out the comparison. The pilot symbols that have been sent, spreading sequences, and other known sequences such as preambles and mid-ambles, are some examples of patterns that may be compared in determining whether a PU is present or not.

As an example, an empirical model that utilises two different sensing strategies in the 802.11b wireless local area network (WLAN) frequency band was proposed in [13]. This type of waveform-based sensing exploits the two states of a channel (occupied or free) to identify spectrum opportunities using a priori knowledge of the transmission scheme of the PUs. Another waveform-based sensing technique, which has been implemented in hardware, uses some ultra wide band (UWB) devices in the worldwide interoperability for microwave access (WiMAX) band to detect and avoid WiMAX devices in some specific areas with strict regulatory status [14].

Performance evaluations of waveform-based sensing have been well conducted with respect to interference estimation and have been shown to be quite useful. One advantage that waveform-based detection techniques have over most other techniques for spectrum sensing is that they have short sensing times. However, the major challenge with waveform-based sensing techniques is that they usually have to cope with issues related to synchronisation, which potentially limits their applicability in CRN designs.

3.3.5   Radio Identification Sensing Techniques

Another technique being used for spectrum sensing in the CRN is by radio identification. In this case for radio identification sensing, the presence of a PU may be detected by simply considering the type of transmission technology that PU is using to transmit its signal. We know that most transmission technologies (Bluetooth, ZigBee and others) have unique features that distinguishe one from another. In the radio identification sensing techniques, feature extraction and classification techniques are used to identify the presence of some known transmission techniques, thereby establishing the presence of PU communication through such technologies [15]. Some of the features of a transmission technology that may be identified and employed to determine the presence of a PU include the amount of energy detected in a signal, the channel bandwidth and centre frequency of a signal or some other statistical information of the signal that is being measured.

With radio identification sensing techniques, the extracted features from a received signal are used to determine and select the most probable PU technology present, and also to determine if that technology can conveniently accommodate the SUs in a possible CRN setup. It has been suggested that radio identification-based sensing may provide improved accuracy over most of the previously mentioned methods of detecting PU signals in the CRN.

3.3.6   Techniques that Employ Multiple Antennas

In performing spectrum sensing for the CRN, the use of multiple antennas is another method that has been investigated in literature. In this case, time and spatial correlation between multiple received signal versions are exploited in determining the presence or absence of a PU [16]. If we assume $M > 1$ receiving antennas and flat fading channel conditions, then some well-known multiple input multiple output (MIMO) techniques may be used for the suboptimal combining of multiple received signals, and in ascertaining the PU's presence in the network [17, 18]. Some examples of MIMO combining techniques that have been well used are the maximum ratio combining (MRC) and equal gain combining (EGC) techniques.

When combining signals in a MIMO design, to help realise the best performance in the spectrum sensing activity for CRN using multiple antennas, the goal should be to maximise the signal-to-noise ratio (SNR) of the combined signal at the receiver end of the network [19]. This goal can be achieved by using the best or most appropriate combining technique for the network design. The challenge of complexity and possible interference are generally associated with the spectrum sensing techniques that require the use of multiple antennas.

The sensing techniques discussed in this section are the most referred to and/or the most employed techniques for sensing spectrum opportunities in possible CRN scenarios. We do not claim that it is an exhausted list as there may be a number of other equally good techniques that have not been mentioned which are being investigated and propagated for achieving spectrum sensing in the CRN. In practical CRN designs, the important goal would be to be able to identify the 'best' technique that suites each need and apply such technique for achieving the desired level of spectrum sensing and accuracy in the CRN.

3.4   Problems Associated with Spectrum Sensing in Cognitive Radio Networks

Depending on the expectations or requirements of the CRN design, any of the techniques already discussed (usually the one deemed to be the most appropriate or most beneficial for the CRN design) can be successfully employed and/or deployed for carrying out spectrum sensing in a typical CRN system. However, none of the individual spectrum sensing techniques guarantees an always accurate outcome. The inaccuracies in the outcome of the spectrum sensing activities are usually as a result of incorrect decisions on the status of the particular band being sensed. The causes of such misjudgements on the status of the frequency bands have been linked to some problems associated with spectrum sensing. The most pronounced problems of spectrum sensing are identified as *the hidden node problem*, *the problem of shadowing*, *the problem of multipath fading* and *the problem of receiver uncertainty* [8, 20].

3.4.1   The Hidden Node Problem

The hidden node problem is one of the problems that could make the spectrum sensing activities in the CRN to be inaccurate. With the hidden node problem, there is a particular node (SU, for instance) in the network that is located at such a position or distance that its sensing range does not cover the location of the PU. This means that the PU is not in the SU's sensing range, and thus, the SU does not 'see' the PU. As a result, the SU is completely unaware of the fact that the PU is transmitting its signals on the PU's channel. The SU therefore senses the absence of a PU on that channel and chooses to transmit its signals, whereas the PU is actually occupying or using the supposedly free channel, causing significant interference to the PU.

3.4.2   The Problem of Shadowing

Shadowing is another problem that could make the spectrum sensing activities in the CRN to be inaccurate. In the shadowing problem, even though the PU is within the sensing range of the SU, the signal transmitted by the PU transmitter does not reach the SU because of the effects of shadowing caused by the high-rise buildings that is in between the PU and the said SU. The authors in [20] discussed the effects of shadowing on spectrum sensing for the CRN. The authors explained that shadowing may cause the SU to be blocked from the PU's signal if there are high-rise buildings between them. The shadowed SU determines that the PU's channel is free, whereas the PU is transmitting its signals on that channel. The SU may then attempt to transmit on the PU's channel, causing significant interference to the PU.

3.4.3   The Problem of Multipath Fading

Multipath fading is one other problem that could make the spectrum sensing activities in the CRN to be inaccurate. With multipath fading, the signals that are transmitted bounces off nearby buildings and vegetations, causing reflection and scattering of such signals. As a result, multiple attenuated versions of the original signals get to the SU, making it difficult for the SU to correctly detect the PU's original signals. The effects of multipath fading on spectrum sensing outcomes were also discussed in [20]. Because of the multipath effect, the SU may struggle in determining correctly whether the signals it receives are from the PU or are from some other sources. It may or may not decide to transmit because of these extraneous signals, making the probability of error in its sensing outcomes higher than it should have normally been.

3.4.4   The Problem of Receiver Uncertainty

The problem of receiver uncertainty could also result in poor sensing outcomes for the CRN. In this case, a particular SU, even though it is located within its own secondary network, is just at the boundary of the range of the PU transmitter. The SU is then unsure of and/or cannot properly determine whether the PU is transmitting or not, thus suffering the problem of receiver uncertainty. The problem with receiver uncertainty is that if the SU chooses to transmit, it could interfere with the PU receiver. The possible effect of receiver uncertainty on spectrum sensing outcomes was also demonstrated in [20]. Due to the SU's position, which is just outside the PU's range, if the SU decides to communicate using that particular channel of the PU, the interference caused to the PU could be beyond the acceptable interference threshold, casting reasonable doubts on the sensing outcomes for the CRN.

3.5   Determining Sensing Accuracy

Given that is it possible for errors to occur in the sensing activities for the CRN, it is imperative to always check the accuracy of the spectrum sensing outcomes and results as they occur. To determine whether the spectrum sensing outcomes in the CRN are okay or not, the accuracy of the results from spectrum sensing activities is usually measured by using a simple binary hypothesis. In the hypothesis, a channel that is unoccupied or free is represented as H_0, while a channel that is occupied is represented as H_1. The parameters H_0 and H_1 are given by:

$$H_0 : a(n) = w(n), \tag{3.1}$$

$$H_1 : a(n) = b(n) + w(n), \tag{3.2}$$

where $a(n)$ is the signal that is measured during the spectrum sensing activity, $b(n)$ is the actual signal that was transmitted and $w(n)$ denotes the additive white Gaussian noise. Therefore, $s(n) = 0$ when a PU is not using a channel. In other words, $|s(n)| > 0$ whenever the PU is available and/or occupying its channel or subchannel.

Using the energy detection technique as an example, we denote the absolute energy measured for $a(n)$ by X. Then, a binary occupancy decision D can be made by comparing X to a noise threshold λ, such that:

$$D = \begin{cases} 1, & X > \lambda \\ 0. & \text{otherwise} \end{cases} \tag{3.3}$$

It is important to always have accurate spectrum sensing results, if the CRN is to achieve the level of performance that is desired for the network. Therefore, spectrum sensing must be carried out with the sole aim of achieving the most accurate outcome possible for the CRN that is designed. Spectrum sensing performances are characterised by two performance measures: the *probability of detection P_d* and the *probability of error P_e*. These measures are defined as follows:

$$P_d = Pr\{X > \lambda \mid H_1\}, \tag{3.4}$$

$$P_e = P_{md} + P_f, \tag{3.5}$$

where $P_{md} = (1 - P_d)$ is the *probability of misdetection* and P_f is the *probability of false alarm*. The probability of sensing a frequency space to be unoccupied and available when it is actually occupied by a PU is referred to as the probability of misdetection. The probability of sensing a channel to be unavailable because it is occupied by a PU when, in actuality, that channel is free or unoccupied by the PU is referred to as the probability of false alarm. P_f is given as:

$$P_f = Pr\{X > \lambda \mid H_0\}. \qquad (3.6)$$

The parameters P_{md}, P_f, P_d and P_e are all critical in the determination of the accuracy of sensing outcomes in the CRN. When performing spectrum sensing, for the best sensing outcomes and results, the goal is always to *minimise P_e* and to *maximise P_d*.

3.6 Cooperative Spectrum Sensing in Cognitive Radio Networks

The problems of possible misjudgements in sensing outcomes in the form of misdetections and false alarms, which are usually caused by the effects of the hidden node, shadowing, multipath, etc., can significantly limit sensing results and render sensing outcomes inaccurate, unreliable and/or unusable. To achieve better sensing outcomes, therefore, *collaborative or cooperative approaches to spectrum sensing* has been proposed to give improved results in sensing outcomes than any and/or all the sensing techniques already discussed. In other words, cooperative sensing techniques are more beneficial in that they are able to achieve better outcomes in the overall spectrum sensing activities than any of the spectrum sensing techniques employed by themselves alone.

3.6.1 Benefits of Cooperative Sensing

In cooperative sensing, by some agreement, the information gathered on the sensing outcomes of each SU are shared with other SUs. As a result, significant reductions in the probabilities of error (misdetection and false alarm) are realised. Furthermore, when cooperative sensing is employed, spectrum sensing results can be improved further by exploiting frequency, space and/or time diversities. The gains as a result of employing cooperative spectrum sensing are mainly experienced in the form of significant improvements in the accuracy of PU signal detection and an increase in the agility of the secondary network.

With cooperative sensing, more than one sensing nodes cooperate to sense the presence of PU signals and/or a free spectrum. Thus, the concepts of *cooperative probability of detection Q_d* and *cooperative probability of false alarm Q_f* have been postulated. Assuming the use of the k-out-of-N fusion rule (this is the most frequently employed hard decision fusion rule for combining multiple units of an entity), the probabilities Q_d and Q_f may be defined as:

$$Q_d = Pr\left\{\sum_{i=1}^{N} D_i \geq k \mid H_1\right\}, \qquad (3.7)$$

$$Q_f = Pr\left\{\sum_{i=1}^{N} D_i \geq k \mid H_0\right\}. \qquad (3.8)$$

In the extreme case where the k-out-of-N rule results in an OR decision (in which case, $k = 1$), Q_d and Q_f become:

$$Q_d = 1 - \prod_{i=1}^{N}(1 - P_{d,i}), \qquad (3.9)$$

$$Q_f = 1 - \prod_{i=1}^{N}(1 - P_{f,i}). \qquad (3.10)$$

Similarly, in the extreme case where the k-out-of-N rule results in an AND decision (in which case, $k = N$), Q_d and Q_f become:

$$Q_d = \prod_{i=1}^{N} P_{d,i}, \qquad (3.11)$$

$$Q_f = \prod_{i=1}^{N} P_{f,i}. \qquad (3.12)$$

The values of the probabilities Q_d and Q_f have been shown to be improved (Q_d is higher and Q_f is lower) when cooperative sensing is employed than by the use of any of the non-cooperative sensing techniques earlier discussed.

3.6.2 The Cost of Cooperative Sensing

The benefits of cooperative sensing are the immediate and definite improvements in the correct detection of PU signals and less misjudgements due to misdectections and false alarms. However, these benefits of cooperative sensing are not without a cost. When cooperative sensing is employed, the complexity of the CRN becomes higher and network algorithms become more challenging to develop and analyse. More so, with cooperative sensing, the process of identifying whether a PU is present or not may result in significant time delays, especially if there is poor synchronisation of the reporting process of the SUs. Again, cooperative sensing requires more energy than non-cooperative sensing. This is because, in cooperative sensing, energy is required not only to drive the process of spectrum sensing by individual SUs, but also for each SU to report the outcomes of their spectrum sensing endeavours for cooperative decisions to be reached on the presence or absence of a PU.

Further, there are usually security concerns when cooperative sensing is employed for spectrum sensing. The security issues associated with cooperation are primarily due to incorrect reporting of local spectrum sensing results by some individual SUs. Generally, SUs that give false reports are either *malicious* or have *malfunctioned*. If not properly handled, they can negatively influence the overall cooperative decision of the CRN system.

3.6.3   Techniques for Cooperative Sensing

In cooperative spectrum sensing for the CRN, two techniques are prominently employed which are the *data fusion* and the *decision fusion* cooperative techniques. In the *data fusion technique*, the SUs cooperate by sending their sensed information to a central processing unit or fusion centre where decisions on the spectrum are made and communicated to each SU in the network [21]. The data fusion technique is a *centralised* cooperative sensing approach. In the *decision fusion technique*, multiple users perform their spectrum sensing *independently* but share their results with other neighbouring SUs. The output from the collective SUs is received by each individual SU and used for their final decision making on whether or not a PU is available and using its channels [22]. The decision fusion technique is a *distributed* cooperative sensing approach.

Finally, decision making during cooperative spectrum sensing can be by *soft combination* (some examples of soft combining are the maximal ratio combining, the optimal combining or the equal gain combining techniques) or by *hard combination* (some examples of hard combining are by the use of linear fusion rules or the majority rules of combining). The soft combination approaches for cooperative spectrum sensing generally achieve better results than the hard combination approaches in decision making activities for the CRN [23].

3.7   Spectrum Prediction for Cognitive Radio Network Applications

As already explained, spectrum sensing is very critical to a successful CRN implementation. In the previous sections, the most promising spectrum sensing techniques for the CRN have already been discussed. However, all the spectrum sensing techniques and approaches (both the non-cooperative and the cooperative techniques discussed) still have some significant limitations and drawbacks that could potentially negatively affect their sensing results.

One of the greatest limitations of spectrum sensing is that *some significant amount of time is spent to perform the spectrum sensing activities*. This reduces the overall time available for use to carry out actual data transmission. Another important drawback of spectrum sensing in the CRN is that, *with spectrum sensing, the overall power consumption of the SUs are increased*. To help improve spectrum sensing outcomes and make the effort well worthwhile, therefore, *new* and *improved* models for spectrum sensing that address the limitations of spectrum sensing are still required.

To address some of the most significant drawbacks of spectrum sensing in the CRN, the concept of *spectrum sensing prediction* has been recently advocated. The main idea of spectrum sensing prediction is that, if it were possible to make accurate predictions about the future activities of the PUs, then we may reduce the total amount of time being spent on spectrum sensing by simply using some of the predicted results of PU activities [8]. Another important advantage of spectrum sensing prediction is that, if the behaviour of PUs are accurately predicted, it will help to make proactive decisions on the channels for the CRN. The process of selecting channels in the CRN can be significantly improved by employing accurate sensing prediction models [24].

In order to make accurate predictions on the future spectrum occupancy of the PUs, the SUs have to first gather sufficient information about the traffic patterns of the PUs. The data generated from the information about the traffic patterns of the PUs can then be used to model or predict the behaviour of the PUs. Predictive models that can be used in predicting PU behaviours can be classified into three categories namely, the *artificial-intelligence-based models*, the *linear models* and the *statistical and moving-average-based models*.

3.7.1   Artificial Intelligence Models

It must be stated that the process of predicting the spectrum occupancy activities of PUs can be a very difficult and complex process. One of the methods that have been proposed to achieve the prediction of the spectrum activities of PUs is by the use of various artificial intelligence techniques. Some artificial intelligence models that have been proposed are *neural networks*, the use of *support vector machines*, *Markov chains* and *hidden Markov models*.

For *neural network models*, artificial neural networks can be employed for spectrum occupancy predictions in the CRN, such as in [25]. An artificial neural network is an adaptive and non-linear model which makes use of an interconnected network of artificial neurons to map an input data to an output data. Specific mathematical or computational models are then used by the neurons to process the information obtained from the mapping exercise. Artificial neural networks have some advantages over other prediction models. For instance, unlike statistical methods, artificial neural networks do not need a priori knowledge of the actual statistical distributions of PU activities in order to make predictions [26]. However, the major problem associated with artificial neural networks is that it can be tedious process to train the model before accurate predictions can be achieved.

The *support vector machine models* perform spectrum prediction in the CRN by combining empirical mode decomposition models with support vector regression models [27]. To perform regression estimations, support vector regression models employ support vector machines and use the principle of structural risk minimisation in determining possible outcomes of such estimations, which can be used in predicting the activities of the PUs.

The *Markov chain* and *hidden Markov models* are designed on the basis that if it is possible to use a sequence of binary states to describe whether a channel is occupied or not, then we can model the presence or absence of a PU on a particular channel by a simple two-state Markov chain [28]. With the two-state model, if no PU is present, the OFF state is activated. If a PU is present, the ON state comes to play. We may then assume that there is a statistical relationship between the PU's past occupancy state, its present state and its future state. Further still, since we know that it is not impossible to record errors in the spectrum sensing outcomes, the probability of getting wrong predictions in the results of the PU state must be incorporated. To achieve this, the Markov chain approach can be extended to using the hidden Markov models. The hidden Markov models ensure that an error probability is assigned to each PU state, thereby mitigating the effects of error in the results of the prediction outcomes [29].

3.7.2   Linear Models

The *linear models* are relatively more simplistic than other models, which gives them a competitive advantage over the other models for predicting PU activities in the CRN. The linear models are based on correlation, auto and linear regressions and least squares. Some linear models that have been employed in predicting PU activities in the CRN are the *normalised least mean square algorithm models* [30, 31], the *logistic regression models* [32], the *linear regression models incorporated with binary time series analysis* [33], the *linear regression models incorporated with correlation analysis* [34], the *autoregression models* [35] and the *recursive least squares algorithm models* [36]. Each of these models have their peculiars characteristics, advantages and shortcomings.

3.7.3 Statistical Models

Statistical models are also being used for predicting PU activities in the CRN [37]. Most statistical methods that have been employed for predicting PU activities in the CRN are usually based on either the *exponential distribution approach* or the *simple nearest neighbour approach*. In the exponential distribution approach, channel occupancy of PUs are modelled using independently exponentially distributed processes of network activities [38]. In the simple nearest neighbour approach, the channel occupancy state of PUs is determined by comparing the measured signal to a predetermined threshold signal value [39].

3.8 Cooperative Spectrum Prediction for Cognitive Radio Network Applications

We have established in a previous section that, for the CRN, cooperative spectrum sensing approaches achieves better sensing results than most other non-cooperative spectrum sensing approaches. A good and logical progression would then be to also consider cooperative spectrum prediction for the CRN. *Cooperative spectrum prediction* would imply that the activities of the PUs can be *predicted collectively or in a cooperative manner* by the SUs in the CRN. The underlying premise for cooperative spectrum prediction is that, since it is possible to improve the spectrum sensing process and results by employing multiple SUs in the sensing activities, then *it is possible to achieve better accuracy in the prediction process and results of the PU activities in the CRN if the SUs also collaborate to carry out the prediction.*

Cooperative spectrum predicting has its challenges too. One main problem that comes up with cooperative spectrum prediction in the CRN is the additional computational complexity of the system due to the predicting process being incorporated. Of course, we understand that accuracy and timeliness are two important qualities that define how successful the prediction outcomes of PU activities are. Consequently, for cooperative spectrum predicting to be practicable for the CRN, models that are computationally less demanding, but yet fairly accurate in the prediction of PU activities, are required. One such model is developed in [8].

The cooperative spectrum prediction model developed in [8] uses both a *forecast engine* and a *fusion centre* to achieve near-accurate prediction of future activities of the PUs. The forecast engine predicts the future patterns and behaviour of the PUs. The fusion centre combines the information from cooperating obtained by the SUs to make the right decisions on the activities of the PUs.

In the model developed in [8], each SUs perform their own spectrum sensing and make their own decisions about the activities of the PUs. Each SU then performs predictions based on the future availability of a channel. Information is collected and collated from the different SUs on their presumed activities of the PUs. The information fusion principle is then used to make a collective decision on the future availability of the channel. The results showed that cooperative spectrum prediction provides greater accuracy than single spectrum prediction. The results are more pronounced at instances when individual SUs have poor channel conditions and had to make predictions on the activities of the PUs under such conditions.

3.9 Practical Examples of Recent Campaign Efforts on Spectrum Availability

We have discussed some of the new and modern attempts by which the outcomes of spectrum sensing endeavours can be improved, particularly by cooperation, prediction or both. These investigations are all geared towards making more and more spectrum available for new technologies. Spectrum improvement is achieved by addressing the problem of spectrum scarcity and revealing spectrum opportunities to help drive newly-evolving wireless networks, such as the CRN.

In this final part of the chapter, we provide some classical examples of measurement campaigns on spectrum usage and availability that have been carried out in recent times. The spectrum campaign examples provided are to buttress all the points that have been clearly articulated, which are *the current inefficiencies in spectrum allocation and usage*, *the potentials that the spectrum offers when such inefficiencies are eradicated*, and *how those potentials can be fully harnessed, especially through new technologies such as the CRN.*

It must be noted that almost all the reports on spectrum measurement campaigns that are available suggest that most of the campaigns have been carried out in the more technologically-advanced countries of the world. The main goal of the various measurement campaigns on the spectrum has been to study the usage patterns of the spectrum, and to discover and expose areas or portions of the spectrum that are being underutilised and, by extension, that could be made available for possible utilisation by the newly-emerging networks such as the CRN. Some good examples of such measurement campaigns around the world on the spectrum and its utilisation patterns, challenges, etc. are found in references [40–43].

Even though there seem to be a fairly sizeable amount of measurement campaigns on the spectrum being carried out in the more technologically-advanced countries of the world, in contrast, the general knowledge on spectrum occupancy in Africa is still very poor. To fill this knowledge gap, a number of campaigns on spectrum occupancy in several parts of Africa are recently being undertaken. Two clear examples of such campaigns in Africa are recent works carried out in the SENTECH Laboratory at the University of Pretoria (this is the authors' research laboratory). These research works have been well reported in references [44] and [45]. These two campaigns are briefly discussed in this section of this chapter.

3.9.1 Spectrum Measurement Campaign on the UHF and GSM Bands

The research project carried out in [44] was specifically to help bridge the wide research gap on spectrum usage measurements in Africa. In the research project, a sizeable amount of information on the degree of occupancy of some spectrum bands in South Africa was collected over a relatively wide frequency rage and, for long time periods, through a mobile autonomous system. Furthermore, the research project developed and used a modular hardware system and a special software environment to measure the degree to which these commercial bands in South Africa are being utilised. The particular frequency bands of interest in the campaign were the ultra-high frequency (UHF) and the global system for mobile communications (GSM) bands. The campaign took sufficient measurements on the UHF (470–854 MHz) band, the GSM 900 downlink (935–960 MHz) band and the GSM 1800 downlink (1805–1880 MHz) band.

A period of 6 weeks was used for the measurement campaign. Over this period of time, the well-developed energy detection technique was employed to carry out continuous data collection in Hatfield, Pretoria, South Africa. The GPS coordinates of the measurement site are S 25^0 $45'$ $11''$ and E 28^0 $13'$ $42''$. The actual site where the measurements were taken is the Hatfield campus of the University of Pretoria. The campus is typically an urban area with a significant amount of student presence. There are lots of office spaces/blocks, sales shops, business premises, primary and secondary schools and student accommodations in the surrounding area of the measurement site.

For the campaign to be successful, noise signals must be properly separated from the actual signals being measured. The authors developed a *maximum normal fit method* to achieve this. This maximum normal fit method developed by the authors extracts the component that carries information in a signal from the component that is noise, and uses this information to calculate how occupied the bands are, or how much of the spectrum is being employed at any given time. The authors then compared the results they gathered on the degree of occupancy of the particular bands that were measured with the results from other measurement campaigns from around the world. Some interesting results of the campaign in comparison with the results of other similar campaigns from around the world were reported and well documented.

One of the interesting results from the measurement campaign was that *only about approximately 20% of the UHF band was being occupied*. However, that percentage of occupancy remained fairly constant and continuous over long periods of time. For the GSM 900 MHz band, *the level of occupancy was at approximately 92%*, a much higher value than the value obtained for the UHF band. For the GSM 1800 band, *the level of occupancy was measured to be at approximately 40%*. From the results of the campaign, a significant difference between the occupancy patterns of the UHF and the GSM bands was observed. While the UHF band had an almost constant occupancy pattern, the occupancy patterns of the GSM 900 MHz and

1800 MHz frequency bands fluctuated, depending on the time of day. Fluctuations of about 10% was recorded for the GSM 900 MHz band, while fluctuations of about 20% was reported for the GSM 1800 MHz band. More so, over the days of the week, slight variations of between 1% and 3% were also reported for the GSM 900 MHz and the GSM 1800 MHz bands, respectively.

The final analysis of the results obtained from the spectrum occupancy efforts of [44] examined how the results compare with similar results of the spectrum occupancy measurements from other parts of the world. The comparison indicates that the occupancy for the UHF and GSM bands in South Africa vary, to some degrees, when compared with the results from other countries. The most important discovery from the campaign is that, in South Africa, *there is a high level of underutilised spectrum*. This is most likely to be true for many other African countries as well, since South Africa is one of the most technologically advanced countries in the continent. Importantly, this underutilised spectrum, if well exploited, can help drive the realisation of new technologies such as the CRN.

3.9.2   Spectrum Measurement Campaign on the TV Broadcast Bands

Most countries in the world have either switched completely or are on the verge of switching over from analogue to digital television (TV) broadcasting. Most countries that are yet to make the complete switch from analogue TV to digital TV do already have target dates by which such complete switch are expected to be completed. A complete switch from analogue to digital TV will provide spectrum regulators the opportunity to accommodate newly emerging technologies such as the CRN. The spectrum bands that will be freed up as a result of this analogue-to-digital switch are called the *TV white spaces* (TVWS). The TVWS will represent the parts (sizeable, of course) of the spectrum in the very high frequency (VHF) and ultra-high frequency (UHF) bands that will become available for use when the analogue-to-digital switch is complete.

The work in [45] studied the potential TVWS opportunities for South Africa. The motivation behind the campaign is the need to discover portions of the spectrum that will become available for potential use by new technologies such as the CRN, when TV becomes fully digitalised in South Africa. In the campaign, three different mechanisms were employed to compare the spectral opportunities that would accrue from TV digitalisation in South Africa. The mechanisms were *some localised spectrum measurements* obtained by the measurement platform developed in [44], *actual spectrum assignments* that was carried out by the South African spectrum regulator [46] and a *locally built and readily available static geolocation database* that was developed from standard wireless propagation models [47].

In the campaign, the authors used the measurement system developed in [44] to take appropriate measurements of the power spectrum of the broadcast TV bands.

The measurements were carried out at six different locations on the Hatfield campus of the University of Pretoria, South Africa. Measurements were taken on the UHF band (470–854 MHz) and the VHF band (174–254 MHz). A frequency resolution of 500 kHz was used to take the measurements. This meant that 500 consecutive time samples were taken in each minute. This was done every minute for a 1 h period.

The report of the campaign in [45] revealed that there are numerous spectrum opportunities through the TVWS. In fact, in some locations, free spectrum spaces worth between 216 and 376 MHz were identified. When the results obtained from the campaign were compared with the results of similar measurement campaigns from other parts of the world, the results from the campaign correlated with the comparative results from similar works in other parts of the world. This strongly validates the campaign process.

3.10 Summary of the Chapter

To summarise our discussions in this chapter, we have shown and emphasised that the spectrum, being an integral component of the CRN, must be well-sourced (or sensed) for optimal opportunities and applications. There are indeed opportunities with the spectrum, and several spectrum measurement campaigns in Africa and other parts of the world confirm this. The concepts of cooperative sensing and predictive sensing are recent but vital tools that will definitely help to make sufficient spectrum available for practical CRN realisations.

References

1. S. Haykin, P. Setoodeh, Cognitive radio networks: the spectrum supply chain paradigm. IEEE Trans. Cogn. Commun. Netw. **1**(1), 3–28 (2015)
2. S. Filin, H. Harada, M. Hasegawa, Performance evaluation of dynamic spectrum assignment and access technologies, in *Proceedings of the IEEE 19th International Symposium on PIMRC* (2008), pp. 1–5
3. J. Pastircak, J. Gazda, D. Kocur, A survey on the spectrum trading in dynamic spectrum access networks, in *Proceedings of the 56th International Symposium on ELMAR* (2014), pp. 1–4
4. J. Mitola, Cognitive radio: an integrated agent architecture for software defined radios. Ph.D. dissertation, KTH (2000)
5. J. Mitola, G.Q. Maguire, Cognitive radio: making software radios more personal. IEEE Pers. Commun. **6**(4), 13–18 (1999)
6. L.E. Doyle, *Essentials of Cognitive Radio*. The Cambridge Wireless Essentials Series, New York (Cambridge University Press, Cambridge, 2009)
7. D.M.M. Plata, Á. Gabriel, A. Reátiga, Evaluation of energy detection for spectrum sensing based on the dynamic selection of detection-threshold, in *Procedia Engineering*, vol. 35 (2012). International Meeting of Electrical Engineering Research 2012, pp. 135–143. http://www.sciencedirect.com/science/article/pii/S1877705812018097
8. S.D. Barnes, A cooperative prediction based approach to spectrum management in cognitive radio networks, Ph.D. Dissertation, University of Pretoria (2016)

9. S. Dannana, B.P. Chapa, G.S. Rao, Spectrum sensing using matched filter detection, in *Intelligent Engineering Informatics*, ed. by V. Bhateja, C.A. Coello Coello, S.C. Satapathy, P.K. Pattnaik (Springer, Singapore, 2018), pp. 497–503

10. Q. Lv, F. Gao, Matched filter based spectrum sensing and power level recognition with multiple antennas, in *2015 IEEE China Summit and International Conference on Signal and Information Processing (ChinaSIP)* (2015), pp. 305–309

11. D. Ghosh, S. Bagchi, Cyclostationary feature detection based spectrum sensing technique of cognitive radio in nakagami-m fading environment, in *Computational Intelligence in Data Mining*, vol. 2, ed. by L.C. Jain, H.S. Behera, J.K. Mandal, D.P. Mohapatra (Springer, New Delhi, 2015), pp. 209–219

12. R. Kishore, C.K. Ramesha, G. Joseph, E. Sangodkar, Waveform and energy based dual stage sensing technique for cognitive radio using RTL-SDR, in *2016 IEEE Annual India Conference (INDICON)* (2016), pp. 1–6

13. S. Geirhofer, L. Tong, B.M. Sadler, A measurement-based model for dynamic spectrum access in WLAN channels, in *MILCOM'06: Proceedings of the 2006 IEEE Conference on Military Communications* (2006), pp. 1–7

14. S.M. Mishra, S. ten Brink, R. Mahadevappa, R.W. Brodersen, Cognitive technology for Ultra-Wideband/WiMax coexistence, in *2007 2nd IEEE International Symposium on New Frontiers in Dynamic Spectrum Access Networks* (2007), pp. 179–186

15. T. Yucek, H. Arslan, A survey of spectrum sensing algorithms for cognitive radio applications. IEEE Commun. Surv. Tutorials **11**(1), 116–130 (2009)

16. J. Ma, G. Zhao, Y. Li, Soft combination and detection for cooperative spectrum sensing in cognitive radio networks. IEEE Trans. Wirel. Commun. **7**(11), 4502–4507 (2008)

17. Y. Liang, Y. Zeng, E.C.Y. Peh, A. T. Hoang, Sensing-throughput tradeoff for cognitive radio networks. IEEE Trans. Wirel. Commun. **7**(4), 1326–1337 (2008)

18. A. Pandharipande J.M.G. Linnartz, Performance analysis of primary user detection in a multiple antenna cognitive radio, in *2007 IEEE International Conference on Communications* (2007), pp. 6482–6486

19. Y. Zeng, Y.-C. Liang, A. Hoang, R. Zhang, A review on spectrum sensing for cognitive radio: challenges and solutions. EURASIP J. Adv. Signal Process. **2010**(1), 381465 (2010). http://asp.eurasipjournals.com/content/2010/1/381465

20. I.F. Akyildiz, B.F. Lo, R. Balakrishnan, Cooperative spectrum sensing in cognitive radio networks: a survey. Phys. Commun. **4**(1), 40–62 (2011). http://www.sciencedirect.com/science/article/pii/S187449071000039X

21. D. Teguig, B. Scheers, V. Le Nir, Data fusion schemes for cooperative spectrum sensing in cognitive radio networks, in *2012 Military Communications and Information Systems Conference (MCC)* (2012), pp. 1–7

22. P. Verma, B. Singh, On the decision fusion for cooperative spectrum sensing in cognitive radio networks. Wirel. Netw. **23**(7), 2253–2262 (2017). https://doi.org/10.1007/s11276-016-1285-0

23. S. Nallagonda, Y.R. Kumar, P. Shilpa, Analysis of hard-decision and soft-data fusion schemes for cooperative spectrum sensing in Rayleigh fading channel, in *2017 IEEE 7th International Advance Computing Conference (IACC)* (2017), pp. 220–225

24. B.S. Shawel, D. Hailemariam Woledegebre, S. Pollin, Deep-learning based cooperative spectrum prediction for cognitive networks, in *2018 International Conference on Information and Communication Technology Convergence (ICTC)* (2018), pp. 133–137

25. Z. Jianli, W. Mingwei, Y. Jinsha, Based on neural network spectrum prediction of cognitive radio, in *2011 International Conference on Electronics, Communications and Control (ICECC)* (2011), pp. 762–765

26. D. Das, D.W. Matolak, S. Das, Spectrum occupancy prediction based on functional link artificial neural network (flann) in ISM band. Neural Comput. Appl. **29**(12), 1363–1376 (2018). https://doi.org/10.1007/s00521-016-2653-5

27. C. Yu, Y. He, T. Quan, Frequency spectrum prediction method based on EMD and SVR, in *2008 Eighth International Conference on Intelligent Systems Design and Applications* **3**, 39–44 (2008)

28. Y. Li, Y. Dong, H. Zhang, H. Zhao, H. Shi, X. Zhao, Spectrum usage prediction based on high-order Markov model for cognitive radio networks, in *2010 10th IEEE International Conference on Computer and Information Technology* (2010), pp. 2784–2788

29. A. Saad, B. Staehle, R. Knorr, Spectrum prediction using hidden Markov models for industrial cognitive radio, in *2016 IEEE 12th International Conference on Wireless and Mobile Computing, Networking and Communications (WiMob)* (2016), pp. 1–7

30. I. Sidi Mohamed Hadj, M. Hachemi, H.E. Adardour, M. Hadjila, Spectrum sensing with VSS-NLMS process in Femto/Macro-cell environments. Int. J. Elect. Comput. Eng. **8**(12), 5185 (2018)

31. X. Tan, H. Zhang, Q. Chen, J. Hu, Opportunistic channel selection based on time series prediction in cognitive radio networks. Trans. Emerg. Telecomm. Technol. **25**(11), 1126–1136 (2014). https://onlinelibrary.wiley.com/doi/abs/10.1002/ett.2664

32. H. Marquez, Prediction of channel availability in cognitive radio networks using a logistic regression algorithm. Int. J. Eng. Technol. **9**(10), 3813–3820 (2017)

33. A. Gorcin, H. Celebi, K.A. Qaraqe, H. Arslan, An autoregressive approach for spectrum occupancy modeling and prediction based on synchronous measurements, in *2011 IEEE 22nd International Symposium on Personal, Indoor and Mobile Radio Communications* (2011), pp. 705–709

34. G.S. Uyanik, B. Canberk, S. Oktug, Predictive spectrum decision mechanisms in cognitive radio networks, in *2012 IEEE Globecom Workshops* (2012), pp. 943–947

35. Z. Wen, T. Luo, W. Xiang, S. Majhi, Y. Ma, Autoregressive spectrum hole prediction model for cognitive radio systems, in *IEEE International Conference on Communications Workshops (ICCW 2008)* (2008), pp. 154–157

36. P. Kulkarni, T. Lewis, Z. Fan, Simple traffic prediction mechanism and its applications in wireless networks. Wirel. Pers. Commun. **59**(2), 261–274 (2011). https://doi.org/10.1007/s11277-009-9916-8

37. H. Eltom, K. Sithamparanathan, R. Evans, Y. Chang Liang, B. Risti, Statistical spectrum occupancy prediction for dynamic spectrum access: a classification. EURASIP J. Wirel. Commun. Netw. **2018**(12), 29 (2018)

38. C. Ghosh, S. Pagadarai, D.P. Agrawal, A.M. Wyglinski, A framework for statistical wireless spectrum occupancy modeling. IEEE Trans. Wirel. Commun. **9**(1), 38–44 (2010)

39. Z. Chen, N. Guo, Z. Hu, R.C. Qiu, Experimental validation of channel state prediction considering delays in practical cognitive radio. IEEE Trans. Vehi. Technol. **60**(4), 1314–1325 (2011)

40. R.I.C. Chiang, G.B. Rowe, K.W. Sowerby, A quantitative analysis of spectral occupancy measurements for cognitive radio, in *2007 IEEE 65th Vehicular Technology Conference – VTC2007-Spring* (2007), pp. 3016–3020

41. M. Lopez-Benitez, F. Casadevall, A. Umbert, J. Perez-Romero, R. Hachemani, J. Palicot, C. Moy, Spectral occupation measurements and blind standard recognition sensor for cognitive radio networks, in *2009 4th International Conference on Cognitive Radio Oriented Wireless Networks and Communications* (2009), pp. 1–9

42. M. Matinmikko, M. Mustonen, M. HÿyhtyÄd', T. Rauma, H. Sarvanko, A. MÄd'mmelÄd', Distributed and directional spectrum occupancy measurements in the 2.4 GHz ISM band, in *2010 7th International Symposium on Wireless Communication Systems* (2010), pp. 676–980

43. T.M. Taher, R.B. Bacchus, K.J. Zdunek, D.A. Roberson, Long-term spectral occupancy findings in chicago, in *2011 IEEE International Symposium on Dynamic Spectrum Access Networks (DySPAN)* (2011), pp. 100–107

44. S. Barnes, P.J. van Vuuren, B. Maharaj, Spectrum occupancy investigation: measurements in South Africa. Measurement **46**(9), 3098–3112 (2013). http://www.sciencedirect.com/science/article/pii/S0263224113002431

45. S. Barnes, P. Botha, B. Maharaj, Spectral occupation of TV broadcast bands: measurement and analysis. Measurement **93**, 272–277 (2016). http://www.sciencedirect.com/science/article/pii/S0263224116303785

46. ICASA, Draft terrestrial broadcasting frequency plan 2013 (2013). Government Gazette, Republic of South Africa 574 (36321)
47. CSIR, Tv white space database (2014). http://whitespaces.meraka.org.za

Part II
Resources to Drive Cognitive Radio Networks

Even though the spectrum has been rightly adjudged to be the most significant resource for cognitive radio networks, there are several other resources that are almost equally essential for an effective and efficient cognitive radio network realisation. The optimal allocation and usage of the spectrum, in combination with all the other network resources for cognitive radio networks, form the important backbone on which this emerging technology can thrive.

Chapter 4
Resource Optimisation Problems in Cognitive Radio Networks

4.1 Complementary Resources for Cognitive Radio Network Applications

In the previous chapter, the spectrum was identified as the most important resource for the successful rollout of most of the modern and newly-evolving next-generation (xG) wireless technologies, especially the cognitive radio networks (CRN). However, the spectrum is not the only and/or exclusive resource on which these technologies depend. In other words, alongside the spectrum resource, *there are a number of other very important resources* that have to be considered for a successful CRN implementation. The mostly-referred and highly-used resources for the CRN, alongside the spectrum or frequency band, are the *bandwidth*, *modulation schemes*, *subchannels or subcarriers, time slots, transmission power* and the *bit or data rates*.

Thus, apart from the spectrum, the above-mentioned resources for the CRN are equally needed for and used up in CRN applications. These network resources, alongside the spectrum, jointly form the strong backbone that supports the operations of the CRN. A CRN system may still not be optimally operated if one or more of these other resources are not available or if they are poorly administered, despite the presence of the requisite spectrum resource. Hence, the appropriate CRN models must *jointly consider these resources, alongside the spectrum resource*, to be able to fully study, analyse and implement the CRN.

Just like the spectrum, the other resources in the CRN are also scarce, limited and generally unavailable. Hence, when developing the CRN, mechanisms by which the limited and scarce resources of the CRN would be assigned, administered, appropriated or allocated so as to realise the utmost productivity must be incorporated in the design of the CRN. In essence, therefore, the discussion in this chapter on resource allocation, administration and management in the CRN is very crucial for its development, its operation and its eventual rollout.

© The Author(s), under exclusive license to Springer Nature Switzerland AG 2022
B. TJ Maharaj, B. S. Awoyemi, *Developments in Cognitive Radio Networks*,
https://doi.org/10.1007/978-3-030-64653-0_4

4.2 Resource Allocation in Cognitive Radio Networks

The concept of *resource allocation (RA) in CRN* covers all aspects of resource (spectrum and others) sharing, distribution or administration for optimal productivity in the CRN. Actually, RA is not exclusive to the CRN. In reality, the concept of RA has been a core part in the design and application of most wireless communication networks. For instance, the concept of RA has been actively researched for orthogonal frequency division multiple access (OFDMA) wireless networks, such as the 4G and LTE-Advanced networks, and for several other conventional and currently operational wireless communication systems.

Since the OFDMA and its newer variants are being actively considered as viable techniques for the CRN, it is important to carefully examine RA concepts and models for OFDMA-based systems. There are already a good number of useful research works on RA in OFDMA-based networks. Readers with keen interest on RA for OFDMA-based systems may consult these references for an in-depth exploration of RA in OFDMA networks [1–7].

Particularly, the study of RA in the CRN is a very critical area of research. In fact, the study of RA is much more important in the CRN than in many other types of wireless communication. The reason is that, unlike in most other wireless communication designs, in the CRN, a primary network must work alongside with a secondary network. Both networks depend on the scarce network resources to drive their communication process. As a result, the problem of resource scarcity in the CRN is much more pronounced than it is in most other wireless communication systems. To reinstate this fact, the important point being stressed is that, in the CRN, both the primary network and the secondary network have some form of right or access to the already scarce resources for the system, making the resources in the CRN to be much more limited and very problematic to allocate or share.

4.3 Resource Allocation Problems in Cognitive Radio Networks

As previously mentioned, when we talk about RA in wireless communication systems, we are essentially talking about the means by which the limited or scarce resources of a particular wireless communication network of interest can be well administered to achieve optimal results. Again, RA problems are very common problems and they are not peculiar to the CRN. For the CRN, particularly, 'RA problems in the CRN' are the problems that arise in the process of seeking the means to *best allocate* the *limited and scarce resources* of the CRN *fairly and favourably* to all the primary and secondary network users or devices, in order to achieve the desired goals of the CRN. The important resources of the CRN have already been identified as the spectrum, subchannels, time slots, bit rates, frequency band, modulation schemes, transmission power, bandwidth, data rates and others.

To help describe the RA problems in the CRN, we again consider how RA problems in wireless communication systems have been described. Generally, RA problems in wireless communication systems, such as the OFDMA-based networks, have been classified into two main groups, which are the *rate adaptive resource allocation* (RARA) problems and the *margin adaptive resource allocation* (MARA) problems [4]. When a RA problem is developed as a RARA problem, in most occasions, the objective is to maximise a given function of the throughput, data rates, fairness, etc. of the primary users (PUs) and/or the secondary users (SUs) in the CRN, with a limited maximum transmission power at the primary and/or secondary base station(s) of the CRN. There are a good number of researchers that have employed the RARA approach to develop and solve their RA problems in wireless communication networks. Some important examples are [8–10].

When a RA problem is developed as a MARA problem, in most instances, the objective is usually to minimise the total transmission power of the network, all the while making sure that the demands of all primary and secondary network users in terms of data rates, throughput, fairness, etc. are all met. Also, there are a good number of researchers that have already employed the MARA approach to develop and solve their RA problems in wireless communication networks. Some important examples are [11–13].

In recent times, there have been useful *adaptations* of both the RARA and MARA approaches that were originally developed to address RA problems in wireless communication systems to now help in analysing and addressing RA problems in the CRN as well [14, 15]. By adopting these approaches, it makes it much easier to classify RA problems in the CRN. If this classification approach is well adopted for the CRN, it implies that we may broadly classify RA problems in the CRN into two categories, namely RARA-CRN problems and MARA-CRN problems.

While the adaptation of the RARA and MARA classifications of the RA problems in the CRN is a good development for studying and analysing the problems, it must be noted, however, that the RA problems in the CRN are a lot more complex and difficult to analyse and solve than in most other wireless communication systems, such as the OFDMA-based networks. Several useful reasons can be given to buttress this position. The first and very significant reason for the unusual difficulty in the RARA-CRN and MARA-CRN problems is that the spectrum, bandwidth and frequency bands that are available for use in the CRN are never constant, but rather they are always fluctuating [16].

The second and equally important reason for the unusual difficulty in the RARA-CRN and MARA-CRN problems is the extra complexities involved in considering the CRN as a heterogeneous system. Even though it is the more realistic CRN consideration, designing the CRN to be heterogeneous poses a great deal of challenge. In such heterogeneous CRN designs, the communication infrastructure should be able to service more than one user or user categories at the same time, on different channels, using different standards and technologies [17, 18]. This makes the CRN to become very complex and difficult to analyse, study or implement.

The third reason for the unusual difficulty in the RARA-CRN and MARA-CRN problems is the challenge of interference. Indeed, there has to be interference issues in the CRN, since both the primary and the secondary networks have to transmit their data, sometimes at the same time. The CRN may lack in productivity if the interference threshold of the primary users (PUs) in the network is unbearably high. More so, there is the possibility of interference among the SUs themselves. It must be stressed that limitation due to interference is one of the greatest challenge the CRN faces in its bid to achieving its promise of great resourcefulness and optimal productivity, and of being the preferred new wireless communication paradigm.

As a result of the possible challenges with the RARA-CRN and MARA-CRN classifications mentioned above, it is imperative to properly investigate and study the basic principles to be employed while adopting and/or adapting the approaches of RARA and MARA used in other wireless communications to the CRN. Thankfully, there are recent and/or ongoing efforts in this regard. The recent developments or studies on MARA and RARA for CRN (for instance, the ones provided in some chapters of this book) are very useful in the evolution of the CRN. This is because they help to ascertain the suitability and applicability of the MARA and RARA approaches for solving the RA problems in the CRN.

4.4   Resource Allocation in Cognitive Radio Networks as Optimisation Problems

It is important to properly define and describe the RA problems in the CRN if we are to be able to adequately analyse and solve them. What stands out in most cases of RA problems in the CRN is the fact that they are usually *optimisation problems*. Therefore, it is important to have a sound understanding of *optimisation* or *programming* if we are to be able to develop the right solution models and to provide proper analysis of the solutions derived for the RA problems in the CRN.

Without doubt, optimisation is a very potent tool that can be employed and explored for solving RA problems in the CRN. Optimisation is an old and powerful analytical tool that has been developed and employed for solving different kinds of problems in the fields of sciences, engineering and technology. Because of its versatility and dependability, optimisation is still being used extensively today in the fields of operations research, pure and applied mathematics, economics, business and financial management, engineering, technology, etc.

When developing a problem as an optimisation problem, the essence is usually to accomplish at least one *objective* (some optimisation problems do have more than one objective to be achieved). The objective(s) can either be the *maximisation* or *minimisation* of a parameter, an entity or a number of entities. An *objective function* is used to capture and describe the objective(s) that the optimisation problem seeks to achieve. Then, before the objectives(s) can be fully realised, one or more limiting *constraints* have to be considered and overcome.

To arrive at viable solutions to optimisation problems that have been developed, care must be taken so as not to violate the constraints; otherwise, the solutions that are obtained for such problems are not going to remain valid. The *decision variables* are a set of components that helps complete the optimisation problems. The decision variables are the parameters to be calculated and obtained when an optimisation problem is being solved. Generally, the solutions obtained for optimisation problems are either optimal or suboptimal solutions. Further discussions and details on optimisation are not provided in this book. For readers who are interested in understanding the preliminaries on optimisation, we suggest these volumes to help provide the needed foundational or fundamental knowledge on optimisation [19–22].

4.5 A General Representation of the Resource Allocation Problems in Cognitive Radio Networks

In this section, a general representation of the formulation of RA optimisation problems in the CRN is presented. This formulation provides a good description of how to represent the objective functions, the constraints and the decision variables in typical RA scenarios for the CRN and the possible interplay between them. The general mathematical representation of the RA problem formulation for the CRN presented in this section follows the basic RA optimisation formulation described in [23].

The authors in [23] established that the RA optimisation problems in the CRN have generally been developed or formulated using the mathematical programming concept known as *integer programming* (IP). This IP area of optimisation is a well-developed branch of programming or optimisation that is mostly used for selecting the most appropriate integer variables to help arrive at optimal solutions to some particular optimisation problems. There are some variants of the IP branch of optimisation that are also very relevant to the development of RA optimisation problem formulations for the CRN, and for solving such RA problems. Some of the variants of IP that are most relevant to the RA problems in the CRN are the *integer linear programming* (ILP), the *integer non-linear programming* (INLP), the *binary integer linear programming* (BILP), the *mixed integer linear programming* (MILP) and the *mixed integer non-linear programming* (MINLP) approaches.

To represent a typical RA optimisation problem for the CRN, we define two vectors $\mathbf{m}$ and $\mathbf{n}$ to have dimensions x and y, respectively. In the context of RA optimisation for the CRN being discussed, the vector $\mathbf{m}$ could represent the set of *transmission power allocations* for the SUs in the network, while the vector $\mathbf{n}$ could represent indicators of the *subchannel allocation*, which, in that case, would then be zero-one variables. Also, we define the set of positive integers $I = \{0, 1, 2, \ldots\}$. Assume that our goal or objective is to maximise the entire network sum throughput. To achieve this objective, the values of $\mathbf{m}$ and $\mathbf{n}$ that maximises the function $f(\mathbf{m}, \mathbf{n})$, must be obtained. This must be achieved with consideration of the constraints $g_i(\mathbf{m}, \mathbf{n}) \leq a_i$, $i = 1, 2, \ldots, r$. All the variable are non-negative. A mathematical representation of the RA problem becomes:

$$\max Q = f(\mathbf{m}, \mathbf{n}) \tag{4.1}$$

subject to

$$g_i(\mathbf{m}, \mathbf{n}) \leq a_i, i = 1, 2, \ldots, d, \tag{4.2}$$

$$m_j \geq 0, \quad j = 1, 2, \ldots, x, \tag{4.3}$$

$$n_k \in I, \quad k = 1, 2, \ldots, y. \tag{4.4}$$

It is simpler to write Eq. (4.2) as:

$$\mathbf{g}(\mathbf{m}, \mathbf{n}) \leq \mathbf{a},$$

where

$$\mathbf{g}(\mathbf{m}, \mathbf{n}) = \begin{bmatrix} g_1(\mathbf{m}, \mathbf{n}) \\ g_2(\mathbf{m}, \mathbf{n}) \\ \vdots \\ g_d(\mathbf{m}, \mathbf{n}) \end{bmatrix},$$

and $\mathbf{a} = [a_1, a_2, \ldots, n_d]^T$.

It is possible that the RA problem in the CRN is developed as a minimisation and not a maximisation problem. Such cases may occur if, for instance, our objective is to minimise the total transmission power or the total energy consumed by the network. The general RA formulation for the CRN developed above still applies in such cases. What only needs to happen is that the function $Q = f(\mathbf{m}, \mathbf{n})$ should be changed back to a form of maximisation function. This is achieved by negating the original (minimisation) objective function. With such negation, the objective function is now a maximisation function. Simply put, $\max R = -f(\mathbf{m}, \mathbf{n})$.

In the general RA problem formulation for the CRN given in Eqs. (4.1)–(4.4), the objective function is captured in Eq. (4.1), the constraints are captured in Eqs. (4.2)–(4.4), while the decision variables are the components m_j and n_k. Essentially, the RA problem has one objective function (Eq. (4.1)). The resource constraints could be more than one; therefore, d resource constraints are accommodated in Eq. (4.2). The variables x are the non-negative variables, while the variables y are the non-negative integer variables.

We can use the general RA formulation for the CRN given above to describe practical CRN scenarios. Say, we assume that the goal of the CRN was to maximise the entire network sum throughput, subject to some resource constraints. In such practical CRN considerations, therefore, Eq. (4.1) would simply be the function that maximises the network throughput, vector **m** would be the set that indicates the transmission power for all the SUs, vector **n** would be the set that indicates subchannel allocation for each SU (this usually takes binary integer values, that is, 0 or 1) and Eq. (4.2) would be the resource constraints (rate requirement constraint, interference limit constraint, transmit power constraint and/or any other resource constraint, as applicable to the network). In most cases, each of these constraints would be individually represented as different equations in the problem formulation.

The basic ideas presented above have already been employed by a wide range of researchers in formulating and/or describing their RA problems for the CRN. Table 4.1 presents classical examples of some works in which the RA problems for the CRN have been formulated using the ideas from the problem definition, mathematical formulation and optimisation for the CRN, as discussed in this section. Indeed, all the RA problems in Table 4.1 were developed and addressed as optimisation problems. In Table 4.1, the decision variables, constraints and objective function for each of the referenced works are highlighted. Such clarity in problem formulation helps researchers seek and analyse solution models that can achieve the goal of optimising resources for the CRN being investigated.

The examples given in Table 4.1 may not have covered all types of problem formulations on RA in the CRN available in literature. Indeed, it is practically impossible to identify and mention all problems and problem formulation on RA in the CRN in this chapter. However, the examples in Table 4.1 are given to provide the needed context on the critical components of RA problem formulations in the CRN, and how they are set up. The examples in Table 4.1 also show how these essential components of the RA problem formulations all work together so as to realise optimal and/or close-to-optimal solutions for the CRN.

4.6 Unique Characteristics of Resource Allocation Optimisation Problems in Cognitive Radio Networks

The general optimisation formulation provided in the previous section represents how most of the RA problems in the CRN have been formulated. In almost all cases and analyses of the RA problems in the CRN, the RA problems have been accurately described to be *complex, non-deterministic polynomial-time hard* (NP-hard) optimisation problems. Therefore, a careful examination of NP-hard problems would give us a much clearer understanding of the nature and characteristics of the RA problems in the CRN.

An *NP-hard problem* is a problem that requires a non-deterministic algorithm to solve them. If such non-deterministic algorithms exist, the NP-hard problem will be solved in polynomial time. The concept of *determinism*, when applied to optimisation, explains that events do not occur all by themselves without causes, rather, there is always a necessary chain of causation for and event to happen [39]. In other words, before a particular event can occur, there must be a proper

Table 4.1 Description of the basic components of RA problem formulations in the CRN

S/N	Problem definition	Objective function	Main constraints	Decision variables	References
1.	Optimal RA in MIMO-based CRN	Maximising the achievable data rate (or total capacity) of SUs	Transmit power limit of SUs, interference limit to PUs, total transmission time of SUs must be equal to the time slot duration	Number of SUs served	[24–27]
2.	Efficient RA for CRN with cooperation	Maximising the sum rate of all SUs	Transmission power budget of the SUs and the relays, interference to PUs within its tolerable threshold, each subchannel can only be allocated to one SU	Achievable rate over a subchannel, power allocated to each subchannel, integer variables of time slot and a binary allocation indicator	[28–30]
3.	Energy-efficient RA for CRN with imperfect sensing and/or femtocells	Maximising bandwidth capacity for SUs	Power constraint on SUs network, minimum rate guarantee for some SUs and best effort service for the remaining SUs, each subchannel can only be allocated to one SU	Transmit power for each SU, the binary channel allocation indicator	[31–33]
4.	Optimal RA in MIMO-OFDMA based CRN	Maximising sum throughput of the SUs	Interference leakage to PUs always below a threshold, each SU must achieve the minimum required data rate, total transmit power of all SUs must be below the available power at base station, no more than one SU is allocated to each subchannel	Data rate on each subchannel	[34–37]
5.	RA for CRN with opportunistic access	Minimising the symbol error rate of the SUs' network transmission	Constraint on the maximum individual power of each SU, a minimum number of symbols must be sent within a time frame, constraint on the minimum acceptable throughput of the network, interference power to PUs must be below a certain threshold	Total power available to the system, transmitted symbol time	[38]

consideration of the set of actions (such actions are usually finite and manageable) that will help drive such an event. The fact that NP-hard problems are said to be 'non-deterministic', therefore, means that investigating possible solutions to those problems is not restricted to one set of actions.

Indeed, there may be a number of actions or a combination of actions that may need to happen to arrive at a correct solution to a particular NP-hard problem. Therefore, there is no need to select a predetermined choice of actions for the solution algorithms, by imposing on them some particular values, parameters or states that limit their choice. Instead, at every solution attempt, the solution algorithms must be allowed to make their choice of the best actions among several opportunities that are possible for solving the particular problem at that time. What this implies is that, it is very possible that, when using a solution algorithm to solve a given problem, even though the same input parameters are used, multiple solutions at different solution attempts may be obtained.

It is necessary to use non-deterministic algorithms while solving NP-hard problems. This is because, non-deterministic algorithms are capable of making important guesses at crucial points in their operation to help their cause of action towards achieving solutions to the problem at hand [40, 41]. Importantly, if the non-deterministic algorithms are correct in their guesses at those crucial points, appropriate solutions to those problems are easily obtained. The other important part of an NP-hard problem is the polynomial time part. The meaning of 'polynomial time' in an NP-hard problem is simply that, if the solution algorithm being employed is a non-deterministic one, and it makes correct and timely guesses, then the time it will take for the algorithm to solve a particular problem is generally bounded by a polynomial.

4.7   Useful Observations on the Resource Allocation Optimisation Problems in Cognitive Radio Networks

From the description of NP-hard problems provided in the previous section, one may infer that it will be quite difficult, though not impossible, to obtain good solutions to the NP-hard optimisation problems in the CRN. Furthermore, the non-determinism of the solution algorithms creates uncertainties in the amount of time required to solve these NP-hard problems. What this means, therefore, is that it may take more time to solve a problem than the time it takes to develop and describe the problem. Without any doubt, the issue of time and timing can be critical for NP-hard problems and solutions.

What we do understand is that, for solutions to be useful and meaningful, they have to be arrived at in good time, especially in modern or emerging wireless communication prototypes such as the CRN. Long-delayed solutions are unacceptable because of the quick changing nature of such networks. If solutions are delayed, the initial conditions upon which a problem is designed may have been significantly altered, and the solutions obtained would no longer be applicable. An ongoing challenge for the CRN is that of developing generalised RA solution models that arrive at solutions very quickly and are not too complex to implement. If such solution methods and/or models are achieved for the RA problems in the CRN, this will definitely go a long way in the implementation and eventual rollout of the CRN.

4.8 Summary of the Chapter

In this chapter, we have been able to establish the important fact that RA problems in the CRN are indeed optimisation problems. Furthermore, we showed that, because of the characteristics of the RA optimisation problems in the CRN (they are usually NP-hard optimisation problems), it is generally very difficult to obtain solutions for them. Therefore, as a matter of urgency, new and/or improved methods or approaches for achieving viable solutions for these problems, with adequate consideration of the peculiar characteristics and limitations of the CRN, must be investigated and delivered.

More so, since it is still an active research space, continuous studies on RA problems and solution approaches for the CRN is pivotal to helping the CRN achieve its goals. Importantly, the new studies on RA problems and solutions for the modern CRN design must consider and identify the key components of the CRN that have been oversimplified or completely ignored by previous researchers, and determine the impact of those omissions or commissions on the results so far provided. After identifying the pros and cons of previous solution attempts, approaches and models, the new and improved RA models must be developed and designed. These new RA models should incorporate all the critical aspects and needs of the evolving CRN, and they should seek to overcome its many limitations. The benefits of the newly-developed RA models for the CRN must be clear in that, when compared with older models, they must show significant improvements in the productivity and performance of the CRN with the incorporation of the new ideas being thought about and implemented.

References

1. Z. Mao, X. Wang, Efficient optimal and suboptimal radio resource allocation in OFDMA system. IEEE Trans. Wirel. Commun. **7**(2), 440–445 (2008)
2. C. Turgu, C. Toker, A low complexity resource allocation algorithm for OFDMA systems, in *Proceedings of the 15th IEEE Workshop on SSP* (2009), pp. 689–692

3. C. Shi, Y. Wang, P. Zhang, Joint spectrum sensing and resource allocation for multi-band cognitive radio systems with heterogeneous services, in *Proceedings of the IEEE GLOBECOM* (2012), pp. 1180–1185

4. X. Yu, T. Lv, P. Chang, Y. Li, Enhanced efficient optimal and suboptimal radio resource allocation in OFDMA system, in *Proceedings of the 6th International Conference on WiCOM* (2010), pp. 1–4

5. S. Kim, B.G. Lee, D. Park, Energy-per-bit minimized radio resource allocation in heterogeneous networks. IEEE Trans. Wirel. Commun. **13**(4), 1862–1873 (2014)

6. S. Bashar, Z. Ding, Admission control and resource allocation in a heterogeneous OFDMA wireless network. IEEE Trans. Wirel. Commun. **8**(8), 4200–4210 (2009)

7. T. Villa, R. Merz, R. Knopp, Dynamic resource allocation in heterogeneous networks, in *Proceedings of the IEEE GLOBECOM* (2013), pp. 1915–1920

8. E.B. Rodrigues, F. Casadevall, Rate adaptive resource allocation with fairness control for OFDMA networks, in *Proceedings of the 18th EW Conference* (2012), pp. 1–8

9. M. Fang, G. Song, Adaptive resource allocation schemes for OFDMA systems with proportional rate constraint, in *Proceedings of the Symposium on CIICT* (2012), pp. 106–110

10. S. Cicalo, V. Tralli, Adaptive resource allocation with proportional rate constraints for uplink SC-FDMA systems. IEEE Commun. Lett. **18**(8), 1419–1422 (2014)

11. H. Liming, X. Lin, Margin adaptive resource allocation with long-term rate fairness considered in downlink OFDMA systems, in *Proceedings of the IEEE EUROCON* (2009), pp. 1919–1923

12. N. Ul Hassan, M. Assaad, Low complexity margin adaptive resource allocation in downlink MIMO-OFDMA system. IEEE Trans. Wirel. Commun. **8**(7), 3365–3371 (2009)

13. M. Pischella, J.-C. Belfiore, Distributed margin adaptive resource allocation in MIMO OFDMA networks. IEEE Trans. Commun. **58**(8), 2371–2380 (2010)

14. B.S. Awoyemi, B.T. Maharaj, A.S. Alfa, QoS provisioning in heterogeneous cognitive radio networks through dynamic resource allocation, in *Proceedings of the IEEE AFRICON* (2015), pp. 1–6

15. B.S. Awoyemi, B.T.J. Maharaj, A.S. Alfa, Solving resource allocation problems in cognitive radio networks: a survey. EURASIP J. Wirel. Commun. Netw. **2016**(1), 176 (2016). https://doi.org/10.1186/s13638-016-0673-6

16. J.-C. Liang, J.-C. Chen, Resource allocation in cognitive radio relay networks. IEEE J. Sel. Areas Commun. **31**(3), 476–488 (2013)

17. Y. Tachwali, F. Basma, H. Refai, Cognitive radio architecture for rapidly deployable heterogeneous wireless networks. IEEE Trans. Consum. Electron. **56**(3), 1426–1432 (2010)

18. B. Awoyemi, B. Maharaj, A. Alfa, Optimal resource allocation solutions for heterogeneous cognitive radio networks. Digital Commun. Netw. **3**(2), 129–139 (2017). http://www.sciencedirect.com/science/article/pii/S2352864816301043

19. W.L. Winston, M. Venkataramanan, *Introduction to Mathematical Programming*, 4th ed. Pacific Grove, London; Thompson Brooks, Cole (2003)

20. P. Pedregal, *Introduction to Optimization*. Texts in Applied Mathematics (Springer, New York, 2004)

21. K. Edwin, H. Stanislaw, *An Introduction to Optimization*, 4th ed. Wiley Series in Discrete Mathematics and Optimization (John Wiley and Sons, Inc., West Sussex, 2013)

22. S. Boyd, L. Vandenberghe, *Convex Optimization*. Berichte über verteilte messysteme (Cambridge University Press, Cambridge, 2004). https://books.google.co.za/books?id=mYm0bLd3fcoC

23. A.S. Alfa, B.T. Maharaj, S. Lall, S. Pal, Mixed-integer programming based techniques for resource allocation in underlay cognitive radio networks: a survey. J. Commun. Netw. **18**(5), 744–761 (2016)

24. M.G. Adian, H. Aghaeinia, Y. Norouzi, Optimal resource allocation for opportunistic spectrum access in heterogeneous MIMO cognitive radio networks. Trans. Emerg. Telecommun. Technol. (2014). http://doi.dx.org/10.1002/ett.2796

25. M.G. Adian, H. Aghaeinia, Optimal resource allocation in heterogeneous MIMO cognitive radio networks. Wirel. Pers. Commun. **76**(1), 23–39 (2014). http://doi.dx.org/10.1007/s11277-013-1486-0

26. M. Adian, H. Aghaeinia, Optimal resource allocation for opportunistic spectrum access in multiple-input multiple-output-orthogonal frequency division multiplexing based cooperative cognitive radio networks. IET Signal Process. **7**(7), 549–557 (2013)

27. M. Adian, H. Aghaeinia, Optimal and sub-optimal resource allocation in multiple-input multiple-output-orthogonal frequency division multiplexing-based multi-relay cooperative cognitive radio networks. IET Commun. **8**(5), 646–657 (2014)

28. S. Wang, M. Ge, C. Wang, Efficient resource allocation for cognitive radio networks with cooperative relays. IEEE J. Sel. Areas Commun. **31**(11), 2432–2441 (2013)

29. S. Wang, Z.-H. Zhou, M. Ge, C. Wang, Resource allocation for heterogeneous cognitive radio networks with imperfect spectrum sensing. IEEE J. Sel. Areas Commun. **31**(3), 464–475 (2013)

30. M. Ge, S. Wang, On the resource allocation for multi-relay cognitive radio systems, in *Proceedings of the IEEE ICC* (2014), pp. 1591–1595

31. R. Xie, F. Yu, H. Ji, Dynamic resource allocation for heterogeneous services in cognitive radio networks with imperfect channel sensing. IEEE Trans. Veh. Technol. **61**(2), 770–780 (2012)

32. R. Xie, F. Yu, H. Ji, Y. Li, Energy-efficient resource allocation for heterogeneous cognitive radio networks with femtocells. IEEE Trans. Wirel. Commun. **11**(11), 3910–3920 (2012)

33. R. Xie, F. Yu, H. Ji, Spectrum sharing and resource allocation for energy-efficient heterogeneous cognitive radio networks with femtocells, in *Proceedings of the IEEE ICC* (2012), pp. 1661–1665

34. Y. Rahulamathavan, S. Lambotharan, C. Toker, A. Gershman, Suboptimal recursive optimisation framework for adaptive resource allocation in spectrum-sharing networks. IET Signal Process. **6**(1), 27–33 (2012)

35. Y. Rahulamathavan, K. Cumanan, L. Musavian, S. Lambotharan, Optimal subcarrier and bit allocation techniques for cognitive radio networks using integer linear programming, in *Proceedings of the 15th IEEE Workshop on SSP* (2009), pp. 293–296

36. Y. Rahulamathavan, K. Cumanan, S. Lambotharan, Optimal resource allocation techniques for MIMO-OFDMA based cognitive radio networks using integer linear programming, in *Proceedings of the 11th IEEE International Workshop on SPAWC* (2010), pp. 1–5

37. Y. Rahulamathavan, K. Cumanan, R. Krishna, S. Lambotharan, Adaptive subcarrier and bit allocation techniques for MIMO-OFDMA based uplink cognitive radio networks, in *Proceedings of the 1st International Workshop on UKIWCWS* (2009), pp. 1–5

38. A. Zafar, M.-S. Alouini, Y. Chen, R. Radaydeh, New resource allocation scheme for cognitive relay networks with opportunistic access, in *Proceedings of the IEEE ICC* (2012), pp. 5603–5607

39. L.E. Doyle, *Essentials of Cognitive Radio*. The Cambridge Wireless Essentials Series (Cambridge University Press, New York, 2009)

40. R.W. Floyd, Nondeterministic algorithms. J. ACM **14**(4), 636–644 (1967). http://doi.acm.org/10.1145/321420.321422

41. O. Tripp, E. Koskinen, M. Sagiv, Turning nondeterminism into parallelism. SIGPLAN Not. **48**(10), 589–604 (2013). http://doi.acm.org/10.1145/2544173.2509533

Chapter 5
Tools for Resource Optimisation in Cognitive Radio Networks

5.1 Solving Resource Allocation Problems in Cognitive Radio Networks

In the previous chapter, we established that the problems of resource allocation (RA) in cognitive radio networks (CRN) are indeed complex problems, and that they may pose a great deal of difficulty when trying to solve them. Thus, being able to adequately, timeously and optimally solve the RA problems that are developed for the CRN is very crucial to its eventual implementation [1]. Already, there are several tools, methods or approaches that have been and are still being developed, investigated and employed to help address or solve these complex non-deterministic polynomial-time (NP)-hard RA problems in the CRN [2].

This chapter examines in depth the various tools or approaches that are being developed and used to solve the RA problems in the CRN. Of course, each tool has its own advantages and disadvantages, and these are well identified and discussed in this chapter. Furthermore, we make good comparison between the tools and draw up useful reports and conclusions on each one of them. We also identified the limitations associated with each solution tool and make useful recommendations on how to address those limitations in order to make those tools usable and productive for the CRN.

A broad classification of the various tools being employed to solve RA problems in the CRN is as follows:

1. The tool of classical optimisation
2. The tool of studying the structure of an RA optimisation problem
3. The tool of the use of heuristics
4. The tool of the use of meta-heuristics.
5. The tool of game theory or multi-objective optimisation
6. The tool of soft computing-based optimisation.

Each one of these tools is critically examined and discussed in details in this chapter.

5.2   The Tool of Classical Optimisation

Classical optimisation is an old, well-developed and very versatile branch or tool of optimisation or programming. Despite its simplicity, classical optimisation still has relevance in modern applications and has been considered and used in solving RA problems in the CRN. Generally, if an RA problem in the CRN falls into any class of classical optimisation, optimal solutions to such a problem can be obtained by using the tool of classical optimisation. One important point with the tool of classical optimisation is that standard solutions already exist to almost all classical optimisation problems. The two most prominent aspects of classical optimisation are *linear programming* (LP) [3] and *convex optimisation* (CO) [4].

In the LP approach of classical optimisation, for a problem to be identified as an LP problem, all the components of that optimisation problem must be linear in nature. The objective function must be a linear function, and all the constraints in the problem formulation must be all linear. There are already a good number of methods that have been well advanced for optimally solving LP problems. Two good examples of such well-established methods for solving LP problems are the simplex method and the interior point method. Therefore, if an RA problem that is developed for the CRN happens to be an LP problem, by employing an appropriate classical LP method, such a problem will surely be solved and optimality will be achieved for the CRN model being investigated.

As an example, the authors in [5] developed a frequency-time allocation problem in cognitive radio wireless mesh network as an LP problem. This problem was solved easily and optimally using the simplex method. Another good example of the use of the simplex method for solving an RA problem in the CRN can be seen in [6]. The problem that was solved was to optimally allocate the frequency bands of the PUs to the SUs in the network. This RA problem was developed and addressed as an LP. The LP solution further demonstrated the region or envelope in which network stability and balance can be achieved for the CRN. In [7], the authors used the interior point method to address the problem of RA (joint transmission, power control and beamforming) for the SUs in the CRN, at instances when the SUs and the PUs are transmitting their data simultaneously.

The CO approach is another classical optimisation approach that has been vastly used for solving complex and modern optimisation problems. For CO, the optimisation problem does not have to be linear before it can be solved. However, such a problem must be shown to be *convex*. In practical applications of CO to solve RA problem in the CRN, non-linear RA problems can still be solved if the convexity of such problems are proven. Also, just as in LP, a good number of methods of

CO have been developed and used to solve CO problems. The *Lagrangian duality* method is a very good example of a CO method that is well developed for solving optimisation problems [8]. The Lagrangian duality approach is mostly applied alongside with the Karush-Kuhn-Tucker (KKT) conditions when employed for the CRN [9].

To be able to successfully employ classical optimisation approaches to optimally solve RA problems in the CRN, the important condition is that such RA problems must *fit into certain structures* of classical optimisation. Such structures are the *linearity* or the *convexity* of optimisation problem. When the RA problems in the CRN have these structures or when they can be modified to have these structures without jeopardising the originality or essence of such problems, the tool of classical optimisation is the best tool for solving such RA problems in the CRN.

As a general observation, most of the classical optimisation methods that are being used to solve the LP or OC RA problems in the CRN are branches of either the simplex method or the interior point method. The following are about the commonest classical optimisation methods, and the corresponding works in which they have been used, to solve RA problems in the CRN: the *barrier* method [10, 11], the *gradient decent* method [12], the *branch-and-bound* method [13, 14], the *lift-and-shift* method [15], the *branch-and-cut* method [16], the *iterative and double-loop iterative* methods [17, 18], the *Lagrangian duality* method [19, 20], the *dual decomposition* method [17, 19] and the *column generation* method [21, 22].

The most significant advantage of the classical optimisation tool and methods for solving RA problems, especially in modern communication networks, is the benefit of obtaining optimal solutions for the RA problems being investigated. Because they achieve optimal solutions, classical optimisation solutions are used as *bounds* for the solutions obtained from using other optimisation tools or methods, which are most times suboptimal. However, the classical optimisation tool also has its own disadvantages or challenges. A major challenge with classical optimisation methods is that it is very difficult to find RA problems in the CRN that just fit nicely into the standard classical optimisation models. This makes it difficult, sometimes impossible, to address and solve such RA problems in the CRN as classical optimisation problems.

Another problem with using classical optimisation to solve RA problems in the CRN is that it is usually difficult, sometimes impossible, to prove that the non-linear programming problems are indeed convex, and that they can be solved as such. The final challenge with classical optimisation being used to solve RA problems in the CRN is that the complexity and computational demands on network resources being used while solving classical optimisation problems can be very high. Unfortunately, time demand and other network resources are usually limited in the CRN, making the classical optimisation tool a not-so-promising tool to consider in practical CRN applications.

5.3   The Tool of Studying the Structure of the Resource Allocation Problems

We have already discussed that, in very many cases of RA problems in the CRN, it is often difficult to nicely fit those RA problems into any standard classical optimisation model. This makes it difficult to directly apply the classical optimisation tool to addressing those problems and obtaining solutions for them. This does not, however, mean that such RA problems in the CRN cannot be solved. In fact, there have been a number of other optimisation tools that have been examined and exploited for solving the RA problems in the CRN, and several other tools are still being investigated.

The tool of studying the structure of the RA problem is one important tool that has been considered for solving RA problems in the CRN. This tool depends on carrying out a careful study of the structure of an RA problem to determine special feature(s) in the problem that can be exploited to either reduce the complexity of the problem or to make the RA problem fit into a classical optimisation model. The resulting problem may be very close to the original problem, if the restructuring process is well done. As such, the solutions obtained from solving the restructured problem may approach optimal, or at least, be close to optimal. There are a number of optimisation methods that employ the study of structure of the problem to solve RA optimisation problems. These optimisation methods are briefly discussed.

5.3.1   The Method of Separation or Decomposition

In RA problems developed for the CRN, it is sometimes possible to *separate* an RA problem into two (or more) simpler problems, while still not destroying the main property or characteristics of the main problem. In other words, the structure of a given RA problem may be studied and exploited to successfully separate or decompose the original problem into some simpler sub-problems. Each sub-problem may then be solved on their own, most times posing a lot less challenge to achieve this. It is usually possible to combine the solutions obtained from the individual sub-problems to obtain a final solution for the RA problem. The combined solution may be the exact solution that would have been obtained if the original problem was solved by itself, or the combined solution may not be the exact solution of the original problem. Even if the solution obtained through separation or decomposition do not exactly equal the solution that would have been obtained by solving the problem directly, the solution through separation may still be useful if it is sufficiently close to the actual solution of the original problem.

There are a number of separation or decomposition methods being employed to solve RA problems in the CRN, such as the Dantzig-Wolfe decomposition method [23]. Some good examples of the use of the method of decomposition for solving RA problems in the CRN are the works in [24] and [17]. The authors in [24], for instance, used the primal-dual decomposition method to obtain optimal solutions to the RA problem being investigated for the CRN. In the analysis of the RA problem, the main problem is split into a number of sub-problems of power allocation for individual users. The sub-problems are easily solved for every decision variable pair, and individual solutions are combined to achieve the final solution for the RA problem in the CRN model being investigated.

In [17], the authors developed a joint spectrum-power allocation model for multiband CRN. To analyse the model, the RA problem was separated into two parts and solved using an iterative dual decomposition method. The work in [25] investigated a decomposition method for solving RA problem in the CRN. The goal of the RA solution was to maximise the utility of the CRN. The RA problem was split into three sub-problems. The first sub-problem was to optimise the assignment of the signal-to-interference-and-noise ratio (SINR), the second sub-problem was to optimise the transmission power of the CRN, while the third sub-problem was to optimise the interference temperature of the secondary network devices in the CRN.

In [22], the authors addressed the combined problem of spectrum sensing, power allocation and channel assignment in cellular CRN by the use of the decomposition method. The original RA problem was developed as a mixed integer non-linear programming (MINLP) problem. However, the RA problem was separated into two sub-problems. The first sub-problem was to optimally sense the spectrum that is available for use in the CRN. The second sub-problem was to optimally allocate the channel and power to all SUs in the CRN. The two sub-problems were solved optimally, meaning that the optimality of the CRN was not sacrificed by the decomposition process.

The most important benefit of the separation or decomposition method for solving RA problems in the CRN is that it is very possible to achieve optimal or very close-to-optimal solutions for the RA problem, usually at a much reduced computational demand. One of the major challenges with the decomposition method, especially when applied to the CRN, is that it is not always possible to decompose all RA problems in the CRN. Another challenge with the decomposition method is that, a good number of RA problems in the CRN will not retain their composition or import when they are separated into smaller sub-problems in order to make them easier to solve.

5.3.2   The Method of Relaxation

As well established, the RA problems in the CRN are usually complex and may be very difficult to solve them. In some cases, the complexity and subsequent difficulty in solving the RA problems developed for the CRN may be because of the presence of an integer constraint in the problem formulation. Integer constraints are common with the CRN since it deals with channel (or subchannel) allocation to network users or devices. Allocating subchannels are usually binary decisions. A subchannel is either allocated to a particular user or it is not allocated to that user. If the subchannel is allocated to a particular user, it takes the value of 1 for that user. If the subchannel is not allocated to that particular user, it takes the value of 0 for that user.

The RA problems in the CRN that have binary integer constraints can be addressed more easily if the integer constraint is relaxed. The relaxation of the binary integer constraint means that the decision variable is no longer imposed to be either 0 or 1. Rather, the decision variable may take any value between 0 and 1. We may then round up or round down the resulting values of the decision variable to get approximate solutions to the RA problem for the CRN that is being considered.

The work in [26] is an example of the use of the method of relaxation for solving RA problems in the CRN. The RA problem in the CRN was initially developed as a mixed integer non-linear programming problem. However, the problem was changed to an LP problem by simply relaxing the integer constraint. This made it possible for the problem to be solved more easily. Another example of the works that have used the method of relaxation to solve their RA problems for the CRN is the work in [27].

The main benefit of the method of relaxation for solving RA problems in the CRN is that it reduces the complexity of the RA problem very significantly, making it easier to arrive at solutions. The disadvantage of the method of relaxation for solving RA problems in the CRN is that they do not give optimal solutions for the RA problems being solved. Furthermore, there may be a significant gap between the solution obtained after relaxation and the optimal solution to the RA problem, which may be undesirable for practical CRN applications.

5.3.3 The Method of Linearisation

While the LP is surely a potent tool for solving RA problems in the CRN, in most cases, the original RA problem developed for the CRN is hardly ever linear in their composition. In many RA problems for the CRN, the objective functions are not linear. Even in cases when the objective functions are linear, there may be one or more constraints that are non-linear. An optimisation problem cannot be treated as an LP problem once the linearity of either the objective function or any of the constraints is difficult to prove.

One important method that can be used to solve non-linear RA problems in the CRN is the method of linearisation. In this method of optimisation, attempt is made to linearise the non-linear part or expression of either the objective function or any of the constraints in the RA problem. If the linearisation is realised, it becomes pretty straightforward to use the tool of classical optimisation to obtain solutions to the linearised problem. It may happen that the resulting linearised part or expression of the original problem is an approximate of the original part or expression of the initial problem. Still, if the values obtained from the linearisation process are sufficiently close to the values of the original expression, the solutions obtained, though they may be suboptimal, are very useful and profitable.

The works in [28, 29] and [27] are good examples of the application of linearisation methods for solving RA problems in the CRN. The work in [27] used a combination of the methods of linearisation, reformulation and relaxation (the methods of reformulation and relaxation are discussed in latter subsections) to solve their RA problem developed for the CRN. In the part where linearisation was employed, one of the constraints was successfully changed into a linear form. The particular constraint was non-linear because both division and multiplication operations were combined. However, the combined multiplication-division operation was changed to a logarithm function, which has linear expressions. It was possible to maintain the equivalency of the RA problem because of the monotonic characteristic of the logarithm function.

The most important benefit of the linearisation method for solving RA problems in the CRN is that if the linearisation process is successful and the RA problem becomes an LP problem, it is much easier to solve the resulting LP in comparison with the original non-LP problem. The major limitation with the linearisation method is that some of the common functions or expressions in the objective functions or constraints of RA problems in the CRN do not have simple equivalent linear expressions that may be employed in achieving the linearisation.

5.3.4   The Method of Reformulation

The method of reformulation is an important method that has been well used for solving NP-hard RA problems in the CRN. The concept of reformulation simply means to generate an equivalent or nearly-equivalent problem to an original problem. This problem regeneration is normally possible after carefully observing the structure of the original RA problem. There are usually some distinct attributes of the problem that may be exploited to generate a distinct replica of the original problem, while still not losing the most important details in the original problem. In many cases, the regenerated or reformulated problem is an easier version of the original problem. Sometimes, it may even be possible to now use the tool of classical optimisation to solve these reformulated problems.

There are a good number of works that have employed the method of reformulation to solve their RA problems in the CRN. The works in [14, 25, 30–34] are all very good examples. In [25], for instance, the authors developed their RA problem as a utility maximisation problem to help share the spectrum available for CRN operations. The RA problem was complex and non-convex because of the tight coupling between the network interference and the transmission power for the network. To help solve the RA problem, it had to be reformulated. The formulated problem now contained the spectral radius constraint sets. A tuning-free geometrically fast convergent algorithm was used to obtain optimal solutions for the reformulated RA problem.

The work in [30] centred on developing useful algorithms to help make decisions that can optimise the use of radio resources in a heterogeneous cognitive wireless network environment. A significant aspect of the solution process was when the problem of heterogeneous base station selection was reformulated. The reformulated problem simply became a minimum cost-flow problem. The reformulated problem was easily solved at minimal computational demands using directional graphs.

The authors in [14, 32–34] used a similar approach in the reformulation of their RA problems in the CRN. All the RA problems were originally non-linear, non-convex NP-hard problems. However, these RA problems were successfully reformulated into integer linear programming (ILP) problems. The newly generated ILP problems were optimally solved by employing the Branch-and-Bound (BnB) LP tool.

The most important benefit of the use of the method of reformulation for solving RA problems in the CRN is that it is possible to obtain optimal solutions to RA problems that are seemingly difficult, once the reformulation process is successfully carried out. Another advantage of the method of reformulation is that, sometimes, the reformulated problems are a lot less computational demanding than the original RA problems. The major challenge with the method of reformulation is that, in some RA problems of the CRN, it may be very difficult to find that special feature or structure of the original problem that can be exploited to achieve the problem reformulation.

5.3.5 The Method of Approximation

The method of approximation is another useful method for solving RA problems in the CRN. A careful study of the structure of an RA problem for the CRN may reveal that it is indeed a particular function in the problem formulation that makes the problem complex. This problematic function may be in the objective function or in one of the constraints. The function may be responsible for making a problem that should have been linear to become non-linear or a problem that should have been convex to become non-convex. This then makes the entire problem difficult to solve. If it were possible to find an approximate substitute to that problematic function, finding solution to the entire problem becomes much easier.

When the method of approximation is to be employed for solving RA problems in the CRN, care must be taken so that the approximate value or function is always close to the original or initial function being approximated. A number of works have used the method of approximation to obtain solutions to the RA problems developed for either the OFDMA-based networks or the CRN. A good example is the work in [35]. In the work, the authors obtained a piece-wise linear function as a close approximate of their best-effort user utility function. An LP-based cluster allocation algorithm was then used to solve the approximated problem. As a result, they were able to achieve maximum utility for their network.

The greatest benefit of the use of the method of approximation for solving RA problems in the CRN is that, even though it is only suboptimal solutions that can be realised, if the substitutes of the approximated functions are very good, these suboptimal solutions can be very close to the optimal and therefore extremely useful. Another important advantage of the use of the method of approximation is that the computation complexity, analyses of problems and time duration for solving the RA problems are well minimised as a result of the approximation of the difficult functions in the problem formulation.

However, there are some disadvantages with the use of the method of approximation for solving RA problems in the CRN. The first problem with the method of approximation is that the approximate values or functions of the original functions being approximated may now have some extra variables. This usually means that more decision variables will appear in the RA problem for the CRN to be solved. Another disadvantage of the method of approximation is the fact that only suboptimal solutions can be obtained once approximate substitutes of original functions are used in arriving at solutions to the RA problems in the CRN.

5.4 The Tool of Heuristics

Heuristics is one of the commonest tools being employed for solving RA problems in the CRN. Actually, for most RA problems in the CRN, it will be almost impossible to solve them through classical optimisation. For many other RA problems, no matter the special feature that is sought to be employed to make them solvable, solving those RA problems will still not work out. Besides, even if it were possible to employ one or more of the tools and methods that have already been discussed to help solve the RA problems for the CRN, in a good number of cases, it may still happen that the solutions provided, though optimal or near-optimal, would have required a great deal of time that would be impracticable in real-life situations. This means that, as a result of the huge computational demands, such solutions would, to a high degree, be difficult to implement, especially for large, practical networks.

As a result of the huge computational and time demands of most of the solution tools being employed for solving RA problems in the CRN, better and faster solution tools are therefore still required. These tools must be able to obtain solutions for the RA problems at a much faster speed and with less complexity or computational demand. In a lot of cases, a heuristic is usually employed to achieve the expectation of obtaining solutions that are good enough for the RA problem in the CRN, usually at a time frame that is reasonable and workable for practical network implementation.

Heuristics do not employ mathematical, analytical or numerical derivations to solve problems. Rather, they use logic and quick reasoning abilities. Because of this, most heuristics are usually designed to solve specific problems. In solving RA problems in the CRN using heuristics, therefore, the solutions are always problem-specific, and in most cases, they are non-transferable to solve other RA problems in the CRN. Besides, heuristic solutions seldom achieve optimal solutions, they are mostly suboptimal.

There are some benefits that heuristic solutions have over other types of solutions for the RA problems in the CRN. One of the greatest advantages of heuristics is that several RA problems that may not be solvable by any other tool of optimisation may be solved by the use of heuristics. Another advantage is that heuristic solutions are usually obtained at reasonable time frames, much less than the time taken by using other optimisation tools, even when the CRN in consideration is large network. The major disadvantage with the use of heuristics for solving the RA problems in the CRN is that they usually only give suboptimal solutions for the RA problem being solved.

There are several methods of heuristics that have been investigated and applied to solve RA problems in the CRN. Some of these methods are discussed.

5.4.1   Greedy Algorithms

Heuristics that employ the method of greedy algorithms are premised on the idea that, in all situations, the immediate or current best step or line of action or operation must be taken. This is usually without any consideration of the choice that would have provided some better results in some latter parts of the solution process if other steps or actions were decided. Some examples of greedy algorithms are the selective greedy algorithms and the distributed greedy algorithms. Some good examples of works that have used greedy algorithms to solve their RA problems in the CRN are [13, 28, 36–38]. Although the solutions that are obtained by using greedy algorithms may not be optimal, such solutions are obtained in good time, making the method a very useful one.

5.4.2   Water-Filling Schemes

Another method of heuristics that has been developed and employed to solve RA problems in the CRN is the water-filling method. The idea used in the water-filling heuristic, and the many variants of this heuristic, is developed from a very popular problem called the water jug problem. There are a good number of works that have used the water-filling heuristic method to solve their RA problems in the CRN. Some of these examples are in [39–44]. The water-filling schemes have the advantage of ease of development and implementation. They also give results that are quite close to optimal and they are usually less computationally demanding than most other methods of achieving solutions to RA problems in the CRN.

5.4.3   Recursive-Based and Iterative-Based Heuristics

One other method of heuristics that has been employed for solving RA problems in the CRN is the recursive-based and the iterative-based heuristics. In this method of heuristics, resource assignment or allocation to users is carried out in either a recursive or an iterative manner. The recursive heuristics use a structure that is based on the process of selection. On the other hand, the iterative heuristics use a structure repetition. What is important in both methods is that the utility of the network or users is gradually increased until when a new process of recursion or iteration does not bring about any significant improvement in the value of the utility being realised. At such a time, the process is terminated. The works in [11, 32] are good examples of works that have employed the method of recursive and iterative heuristics in solving their RA problems in the CRN.

5.4.4   Pre-assignment and Reassignment Algorithms

Pre-assignment and reassignment algorithms have been used to solve a number of RA problems in the CRN. When pre-assignment and reassignment algorithms are being used, some network resources, say subchannels or transmission power, are initially given as base resources to some or all users in the network. The remaining resources are now allocated to the other users as fairly and optimally as possible. With each run of the algorithm, one or more users are allocated more resources, which increase the overall capacity or productivity of the network. The algorithm checks that the constraints are not violated after each run. After each run, if the algorithm determines that there are some residual resources, it reallocates (or reassigns) such resources to the appropriate users thereby improving the overall productivity of the network. The works in [13, 45] are good examples of the use of the pre-assignment and reassignment method for solving RA problems in the CRN.

5.5   The Tool of Meta-heuristics

The tool of meta-heuristics is an important tool that has been well used for addressing RA problems in the CRN. Meta-heuristics are mostly used when the RA problems in the CRN are very computationally demanding. Meta-heuristics have a broad range of application. They are particularly well suited for solving RA problems in which it is possible to have local 'optimal' solutions. They can also solve RA problems which do not have satisfactory problem-specific algorithms that can solve them.

The important benefit of meta-heuristics is that they can give approximate solutions to any kind of wide range or hard optimisation problem. They also have the advantage of not having to adapt too deeply to each problem before they can solve them [46]. Another important advantage of meta-heuristics is their use of tricks that frees them from a local minima or maxima solution, in order to achieve a possibly better solution. There are a number of methods of meta-heuristics that are being used to solve RA problems in the CRN. Some of these methods are discussed.

5.5.1 Genetic Algorithms

In genetic algorithms, the idea of genetics is being employed to solve RA problems in the CRN. To achieve this, resources are defined in the form of genes and chromosomes. Furthermore, the quality of service requirements for each user or user category are given as the input to the algorithm and used to obtain solutions for the RA problem. One good example of the use of genetic algorithm for solving RA problems in the CRN is the work in [47]. In the work, spectrum allocation in the CRN was optimised by the use of a genetic algorithm. In the work in [48], a genetic algorithm was employed to optimise spectrum utilisation, guaranteeing fairness among the users in the CRN.

5.5.2 Simulated Annealing

The method of simulated annealing is another important meta-heuristic method that has been applied for solving RA problems in the CRN. The simulated annealing method uses the process of continuous 'heating' and 'cooling' of the 'search space' of a problem to solve such a problem. This 'heating' and 'cooling' process is carried out in an iterative manner under strict monitoring and control. Usually, the process produces an optimal 'temperature', which is indicative of the optimal utility for the CRN. The works in [49, 50] employed the simulated annealing method to solve the RA problems of subchannel allocation and utility maximisation in the CRN.

5.5.3 Tabu Searches

The method of Tabu searches is another method of heuristics that has been employed for achieving RA solutions in the CRN. The Tabu search algorithms explicitly use historical results of past searches that have been carried out to arrive at new solutions. They can use these past results to help them escape from a local minima or a local maxima solution that is not optimal. They can also use these past results to implement an explorative strategy to help achieve new results for the network. The important advantage of the use of Tabu searches is that they use mechanisms. that are inspired by the human memory. One good example of the use of Tabu searches in solving RA problems in the CRN is found in the work in [51]. In the work, Tabu searches were employed to achieve optimal channel allocation for all users in the CRN.

5.5.4   Evolutionary Algorithms

Another important method of meta-heuristics that has been well used for solving RA problems in the CRN is the use of evolutionary algorithms. These algorithms attempt to simulate the processes that occurred as certain organisms evolved. The particular processes of interest are the processes of selection, recombination and mutation that happened in the cause of generating better species or solutions. There are already a good number of evolutionary algorithms that are applied to solving RA problems in the CRN, and several more evolutionary algorithms are being developed and implemented for the CRN.

The most commonly used evolutionary algorithms are the *particle swarm optimisation* algorithm, the *coco search* algorithm, the *bee colony* algorithm and the *ant colony* algorithm. An example of a work that used the particle swarm optimisation algorithm to solve a RA problem in the CRN is the work in [52]. In the work, the particle swarm optimisation algorithm was used to carry out power allocation for users in the CRN. The work in [53] employed the bee colony algorithm to carry out relay assignment and power allocation for users in the CRN.

The discussions thus far presented on the use of the tools of heuristics and meta-heuristics for solving RA problems in the CRN show that these tools are very powerful and quite important when solving practical RA problems in the CRN, especially when the networks are substantial. The greatest challenge with the use of heuristics and meta-heuristics for solving RA problems in the CRN is that they lack analytical definitions and numerical representations of the RA problems being solved. This makes it very difficult to transfer the knowledge used or acquired in solving a particular problem to help solve other RA problems in the CRN.

5.6   The Tool of Multi-objective Optimisation and Game Theory

The tool of multi-objective optimisation and game theory is a great tool that is being used to solve RA problems in the CRN. The tool of multi-objective optimisation and game theory is especially useful when the RA problems have multiple objectives. Actually, it is not impossible to have RA problems in the CRN that have more than one objective. In such cases, the resulting RA optimisation problems are multi-objective optimisation problems. Usually, such problems require that two or more possibly conflicting objectives are optimised simultaneously, subject to some defined constraints.

An important approach that has been used to solve a good number of multi-objective optimisation problems in the CRN is to convert them to conventional single-objective optimisation problems. There are a number of conversion methods that are being used to change multi-objective optimisation problems to single-objective optimisation problems. The most common methods are the *Min-Max* method, the *reducing dimension* method, the *interactive programming* method, the *ideal point* method, the *virtual target* method, the *weighted sum of squares* method, the *feasible direction* method, the *centre* method and the *sequencing method* [54].

Even though some of the RA problems in the CRN that are originally developed as multi-objective optimisation problems may be successfully changed to conventional optimisation problems, it is not in all cases that such conversion can take place. Besides, even if the conversion does take place, there are many instances in which, despite the conversial to conventional optimisation problems, solving those multi-objective problems may still be very difficult or impossible using conventional optimisation. In all of those exceptional cases, the use of multi-objective optimisation methods such as game theory is most critical to solving those RA problems in the CRN.

There are several multi-objective optimisation or game theory methods that are being used to solve multi-objective RA problems in the CRN. The most common game theory methods that have been used for solving RA problems in the CRN, and some of the works in which those methods have been employed, are the *Nash bargaining* (or Pareto optimisation) [55, 56], the *Stackelberg game* [57, 58], the *cooperative game* [43, 44] and the *non-cooperative game* [59]. The greatest advantage of the tool of multi-objective optimisation and game theory for solving RA problems in the CRN is that they are very useful for solving those RA problems with multiple objectives in their problem formulation. The major disadvantage of this tool is that they may not give optimal solutions in most use cases.

5.7   The Tool of Soft Computer-Based Optimisation

One of the most recent tools that is being employed to solve RA problems in the CRN is the tool of soft computing-based optimisation. This tool uses software and computer programming to optimally allocate resources among the users and user categories in the CRN. The software and computer programmes that are being developed and used to achieve the RA optimisation employ modern intelligent-based methods to carry out the RA in the CRN. These intelligent methods can learn from their past solution attempts to improve on current and future solutions for the RA problem being investigated. The learning process is referred to either machine learning or deep learning. The most common intelligence-based methods being used to solve RA problems in the CRN are *neural networks*, *artificial intelligence*, *fuzzy systems* method and the *Q-learning* method [60].

There are already few works that have employed the tool of soft computing in solving their RA problems in the CRN. The work in [61] employed a special type of Q-learning, called *multi-agent reinforcement learning*, to achieve RA in a multi-user CRN scenario. In the course of carrying out the learning activity, each SU takes the channel and other SUs to be its learning space. They use their learning experiences to update their Q-values and to make decisions on what they think would be the most viable course of action or activity, usually based on the immediate condition of the network.

The work in [62] developed a powerful tool for making decisions on allocating resources in the CRN using the method of artificial intelligence. In the work, a decision-making engine was proposed using the basic idea of Bayesian network. This engine was used to achieve cognitive radio learning and to obtain optimum configuration rules for the network. The model was able to adapt to the different environmental situations with the learning algorithms that was developed.

The authors in [63] used a fuzzy neural system to carry out spectrum allocation in the CRN. The model developed used some important network parameters such as the degree of mobility of the SUs, the distance of the SUs to the PUs, the spectrum utilisation efficiency of the network, etc. as the input parameters to the fuzzy logic engine to use in making decisions for the network. The output from the fuzzy logic engine is usually the decisions on spectrum access and allocation for the SUs. The fuzzy logic engine used linguistic knowledge of some preconceived rules in reaching its decisions.

The most important benefit of the use of the tool of soft computing-based optimisation for solving RA problems in the CRN is that, since they are based on the use of software and computer programming, they can achieve optimal solutions in reasonable time frames. More so, since they can learn from their past solutions, they can adapt to new environments and challenges and can always improve the current and future solutions they provide to RA problems in the CRN. The major limitation of this tool for solving RA problems in the CRN is that most of the soft computing tools are still works-in-progress, since they are still being actively researched, developed and implemented. Besides, some of these tools are very complex and quite difficult to develop, analyse and implement for practical CRN realisations.

5.8 Summary of the Chapter

This chapter has explored the various tools that have been/are being developed and employed for solving RA problems in the CRN. The tools discussed in this chapter are the most common tools successfully exploited for obtaining solutions to the complex A problems for the CRN. The chapter has presented the classifications of the various solution tools and explained their workability, benefits and challenges. As we conclude, Table 5.1 provides a succinct summary of the distinct characteris-

Table 5.1 A summary of the optimisation tools for solving RA problems in the CRN

S/N	Solution tools	Examples of solution models	Specific features	Drawbacks
1.	Classical optimisation, e.g. LP, convex optimisation, etc.	Simplex and its variants (BnB, BnC, LnS, implicit enumeration, etc.); interior point method and its variants (barrier method, Newton's method, etc.); Lagrangian duality; knapsack; travelling salesman problem, etc.	Approach gives optimal solutions; solutions act as bounds (upper or lower) to other solution models	Usually, most RA problems do not fit into any class of classical optimisation; proving convexity can be very challenging; obtaining solutions can be rather computationally complex and time consuming
2.	Studying problem structure	Decomposition; linearisation; relaxation; approximation; reformulation	Solutions can be optimal or very close to optimal; computational complexity is significantly lowered	Special features might be unavailable or difficult to find; transformed problem may be a far cry from the original; new problem may generate more decision variables than in the original one; solutions are mostly suboptimal
3.	Heuristics	Greedy algorithms; water-filling algorithms; pre-assignment and reassignment algorithms; iterative-based and recursive-based algorithms	Solutions are quick to find; less computational complexity; requires little or no numerical analysis; solutions are usually suboptimal but could be close to optimal; approach is suitable for large and practical networks	Solutions are problem-specific and most times are not transferable; solutions cannot be numerically analysed; solutions are always suboptimal
4.	Meta-heuristics	Genetic algorithms; simulated annealing; evolutionary algorithms; tabu searches	Algorithms are mostly nature-inspired; they make use of stochastic components (e.g. random variables); they are good with large, practical and/or computationally demanding problems that have large search spaces; they use 'tricks' so as not to get stuck at a local optimal but to try obtain a global optimal solution	Solutions are not transferable; solutions cannot be analysed numerically

(continued)

Table 5.1 (continued)

S/N	Solution tools	Examples of solution models	Specific features	Drawbacks
5.	Multi-objective optimisation (using game theory)	Cooperative game; non-cooperative game; Nash bargaining (Pareto optimisation); Stackelberg game	They are good with problems that have multiple objectives; they employ ideas from game theory to solve optimisation problems; they are useful for large, practical networks with large search spaces	Solution models can be complex; they are not transferable; there may be difficulty in achieving analytical modelling of solutions
6.	Soft computing-based optimisation	Artificial intelligence; neural networks; Q-learning; fuzzy systems, etc.	Software/computer-based programming is used in allocating resources to users within the network; the developed programmes use intelligent and very powerful/sophisticated techniques	They are very difficult and complex to develop, analyse and apply in real-life scenarios

tics of each solution tools for RA optimisation in the CRN, as already well discussed in this chapter.

References

1. B. Awoyemi, B. Maharaj, A. Alfa, Optimal resource allocation solutions for heterogeneous cognitive radio networks. Digit. Commun. Netw. **3**(2), 129–139 (2017). http://www.sciencedirect.com/science/article/pii/S2352864816301043
2. B.S. Awoyemi, B.T.J. Maharaj, A.S. Alfa, Solving resource allocation problems in cognitive radio networks: a survey. EURASIP J. Wirel. Commun. Netw. **2016**(1), 176 (2016). https://doi.org/10.1186/s13638-016-0673-6
3. W.L. Winston, M. Venkataramanan, *Introduction to Mathematical Programming*, 4th edn. (Thompson Brooks/Cole, Pacific Grove/London, 2003)
4. S. Boyd, L. Vandenberghe, *Convex Optimization*. ser. Berichte über verteilte Messysteme (Cambridge University Press, Cambridge, 2004). https://books.google.co.za/books?id=mYm0bLd3fcoC
5. G. Zhao, J. Li, K. Lee, J.B. Song, Optimal frequency-time allocation in cognitive radio wireless mesh networks. IETE Techn. Rev. **28**(5), 434–444 (2011). http://www.tandfonline.com/doi/abs/10.4103/0256-4602.85976
6. A. El Shafie, A. Sultan, T. Khattab, Band allocation for cognitive radios with buffered primary and secondary users, in *Proceedings of the IEEE WCNC* (2014), pp. 1508–1513
7. F. Wang, W. Wang, Robust beamforming and power control for multiuser cognitive radio network, in *Proceedings of the IEEE Conference and Exhibition on Global Telecommunications (GLOBECOM)* (2010), pp. 1–5
8. Z.-Q. Luo, W. Yu, An introduction to convex optimization for communications and signal processing. IEEE J. Sel. Areas Commun. **24**(8), 1426–1438 (2006)

9. B.S. Awoyemi, B.T. Maharaj, Mitigating interference in the resource optimisation for heterogeneous cognitive radio networks, in *Proceedings of the IEEE 2nd Wireless Africa Conference (WAC)* (2019), pp. 1–6

10. S. Wang, M. Ge, C. Wang, Efficient resource allocation for cognitive radio networks with cooperative relays. IEEE J. Sel. Areas Commun. **31**(11), 2432–2441 (2013)

11. S. Du, F. Huang, S. Wang, Power allocation for orthogonal frequency division multiplexing-based cognitive radio networks with cooperative relays. IET Commun. **8**(6), 921–929 (2014)

12. W.-C. Pao, Y.-F. Chen, Adaptive gradient-based methods for adaptive power allocation in OFDM-based cognitive radio networks. IEEE Trans. Veh. Technol. **63**(2), 836–848 (2014)

13. Z. Mao, X. Wang, Efficient optimal and suboptimal radio resource allocation in OFDMA system. IEEE Trans. Wirel. Commun. **7**(2), 440–445 (2008)

14. Y. Rahulamathavan, K. Cumanan, S. Lambotharan, Optimal resource allocation techniques for MIMO-OFDMA based cognitive radio networks using integer linear programming, in *Proceedings of the 11th IEEE International Workshop on SPAWC* (2010), pp. 1–5

15. M.Z. Bocus, J.P. Coon, N.C. Canagarajah, J.P. McGeehan, S.M.D. Armour, A. Doufexi, Resource allocation for OFDMA-based cognitive radio networks with application to h.264 scalable video transmission. EURASIP J. Wirel. Commun. Netw. **2011**(1), 245673 (2011). http://jwcn.eurasipjournals.com/content/2011/1/245673

16. M.-S. Cheon, S. Ahmed, F. Al-Khayyal, A branch-reduce-cut algorithm for the global optimization of probabilistically constrained linear programs. Math. Program. **108**(2), 617–634 (2006). http://dx.doi.org/10.1007/s10107-006-0725-5

17. C. Shi, Y. Wang, P. Zhang, Joint spectrum sensing and resource allocation for multi-band cognitive radio systems with heterogeneous services, in *2012 IEEE Global Communications Conference (GLOBECOM)* (2012), pp. 1180–1185

18. S. Kim, B.G. Lee, D. Park, Energy-per-bit minimized radio resource allocation in heterogeneous networks. IEEE Trans. Wirel. Commun. **13**(4), 1862–1873 (2014)

19. L. Wang, W. Xu, Z. He, J. Lin, Algorithms for optimal resource allocation in heterogeneous cognitive radio networks, in *2009 2nd International Conference on Power Electronics and Intelligent Transportation System (PEITS)*, vol. 2 (2009), pp. 396–400

20. R. Xie, F. Yu, H. Ji, Dynamic resource allocation for heterogeneous services in cognitive radio networks with imperfect channel sensing. IEEE Trans. Veh. Technol. **61**(2), 770–780 (2012)

21. J. Zhang, Z. Zhang, H. Luo, A. Huang, A column generation approach for spectrum allocation in cognitive wireless mesh network, in *Proceedings of the 2008 IEEE Global Telecommunications Conference* (2008), pp. 1–5

22. Z. He, S. Mao, S. Kompella, A decomposition approach to quality-driven multiuser video streaming in cellular cognitive radio networks. IEEE Trans. Wirel. Commun. **15**(1), 728–739 (2016)

23. P.L. Vo, D.N.M. Dang, S. Lee, C.S. Hong, Q. Le-Trung, A coalitional game approach for fractional cooperative caching in content-oriented networks. Int. J. Comput. Telecomm. Netw. **77**(C), 144–152 (2015). http://dx.doi.org/10.1016/j.comnet.2014.12.005

24. M. Adian, H. Aghaeinia, Optimal resource allocation for opportunistic spectrum access in multiple-input multiple-output-orthogonal frequency division multiplexing based cooperative cognitive radio networks. IET Signal Process. **7**(7), 549–557 (2013)

25. L. Zheng, C.W. Tan, Cognitive radio network duality and algorithms for utility maximization. IEEE J. Sel. Areas Commun. **31**(3), 500–513 (2013)

26. F. Chen, W. Xu, Y. Guo, J. Lin, M. Chen, Resource allocation in OFDM-based heterogeneous cognitive radio networks with imperfect spectrum sensing and guaranteed QoS, in *2013 8th International Conference on Communications and Networking in China (CHINACOM)* (2013), pp. 46–51

27. P. Li, S. Guo, W. Zhuang, B. Ye, On efficient resource allocation for cognitive and cooperative communications. IEEE J. Sel. Areas Commun. **32**(2), 264–273 (2014)

28. C. Turgu, C. Toker, A low complexity resource allocation algorithm for OFDMA systems, in *2009 IEEE/SP 15th Workshop on Statistical Signal Processing* (2009), pp. 689–692

29. Y. Shi, Y. Hou, A distributed optimization algorithm for multi-hop cognitive radio networks, in *IEEE INFOCOM 2008 - The 27th Conference on Computer Communications* (2008)
30. M. Hasegawa, H. Hirai, K. Nagano, H. Harada, K. Aihara, Optimization for centralized and decentralized cognitive radio networks. Proc. IEEE **102**(4), 574–584 (2014)
31. X. Yu, T. Lv, P. Chang, Y. Li, Enhanced efficient optimal and suboptimal radio resource allocation in OFDMA system, in *2010 6th International Conference on Wireless Communications Networking and Mobile Computing (WiCOM)* (2010), pp. 1–4
32. Y. Rahulamathavan, S. Lambotharan, C. Toker, A. Gershman, Suboptimal recursive optimisation framework for adaptive resource allocation in spectrum-sharing networks. IET Signal Process. **6**(1), 27–33 (2012)
33. Y. Rahulamathavan, K. Cumanan, L. Musavian, S. Lambotharan, Optimal subcarrier and bit allocation techniques for cognitive radio networks using integer linear programming, in *Proceedings of the 15th IEEE Workshop on SSP* (2009), pp. 293–296
34. Y. Rahulamathavan, K. Cumanan, R. Krishna, S. Lambotharan, Adaptive subcarrier and bit allocation techniques for MIMO-OFDMA based uplink cognitive radio networks, in *Proceedings of the 1st International Workshop on UKIWCWS* (2009), pp. 1–5
35. S. Bashar, Z. Ding, Admission control and resource allocation in a heterogeneous OFDMA wireless network. IEEE Trans. Wirel. Commun. **8**(8), 4200–4210 (2009)
36. W. Guo, X. Huang, Maximizing throughput for overlaid cognitive radio networks, in *MILCOM 2009 - 2009 IEEE Military Communications Conference* (2009), pp. 1–7
37. P. Mitran, L.B. Le, C. Rosenberg, Queue-aware resource allocation for downlink OFDMA cognitive radio networks. IEEE Trans. Wirel. Commun. **9**(10), 3100–3111 (2010)
38. E. Driouch, W. Ajib, A. Ben Dhaou, A greedy spectrum sharing algorithm for cognitive radio networks, in *2012 International Conference on Computing, Networking and Communications (ICNC)* (2012), pp. 1010–1014
39. T. Peng, W. Wang, Q. Lu, W. Wang, Subcarrier allocation based on water-filling level in OFDMA-based cognitive radio networks, in *2007 International Conference on Wireless Communications, Networking and Mobile Computing* (2007), 196–199
40. A. Zafar, M.-S. Alouini, Y. Chen, R. Radaydeh, New resource allocation scheme for cognitive relay networks with opportunistic access, in *Proceedings of the IEEE ICC* (2012), pp. 5603–5607
41. Y. Liu, L. Liu, C. Xu, Spectrum underlay-based water-filling algorithm in cognitive radio networks, in *2011 International Conference on Electric Information and Control Engineering* (2011), pp. 2614–2617
42. R. Ujjwal, C. Rai, N. Prakash, Fair adaptive resource allocation algorithm for heterogeneous users in OFDMA system, in *2014 International Conference on Signal Processing and Integrated Networks (SPIN)* (2014), pp. 402–406
43. M.G. Adian, H. Aghaeinia, Y. Norouzi, Optimal resource allocation for opportunistic spectrum access in heterogeneous MIMO cognitive radio networks. Trans. Emerg. Telecommun. Technol. **27**, 74–83 (2014). https://doi.org/10.1002/ett.2796
44. M.G. Adian, H. Aghaeinia, Optimal resource allocation in heterogeneous MIMO cognitive radio networks. Wirel. Pers. Commun. **76**(1), 23–39 (2014). https://doi.org/10.1007/s11277-013-1486-0
45. A. Alshamrani, X. Shen, L.-L. Xie, QoS provisioning for heterogeneous services in cooperative cognitive radio networks. IEEE J. Sel. Areas Commun. **29**(4), 819–830 (2011)
46. I. Boussaid, J. Lepagnot, P. Siarry, A survey on optimization metaheuristics. Inf. Sci. **237**, 82–117 (2013). Prediction, Control and Diagnosis using Advanced Neural Computations. http://www.sciencedirect.com/science/article/pii/S0020025513001588
47. Y. El Morabit, F. Mrabti, E. Abarkan, Spectrum allocation using genetic algorithm in cognitive radio networks, in *2015 Third International Workshop on RFID And Adaptive Wireless Sensor Networks (RAWSN)* (2015), pp. 90–93
48. L. Zhu, Y. Xu, J. Chen, and Z. Li, The design of scheduling algorithm for cognitive radio networks based on genetic algorithm, in *2015 IEEE International Conference on Computational Intelligence & Communication Technology* (2015), pp. 459–464

49. E. Meshkova, J. Riihijarvi, A. Achtzehn, P. Mahonen, Exploring simulated annealing and graphical models for optimization in cognitive wireless networks, in *GLOBECOM 2009-2009 IEEE Global Telecommunications Conference* (2009), pp. 1–8

50. B. Ye, M. Nekovee, A. Pervez, M. Ghavami, TV white space channel allocation with simulated annealing as meta algorithm, in *2012 7th International ICST Conference on Cognitive Radio Oriented Wireless Networks and Communications (CROWNCOM)* (2012), pp. 175–179

51. V. Jayaraj, J. Amalraj, S. Hemalatha, An analysis of genetic algorithm and tabu search algorithm for channel optimization in cognitive adhoc networks. Int. J. Comput. Sci. Mob. Comput. **3**(7), 60–69 (2014)

52. S. Motiian, M. Aghababaie, H. Soltanian-Zadeh, Particle swarm optimization (PSO) of power allocation in cognitive radio systems with interference constraints, in *2011 4th IEEE International Conference on Broadband Network and Multimedia Technology* (2011), pp. 558–562

53. S. Ashrafinia, U. Pareek, M. Naeem, D. Lee, Binary artificial Bee colony for cooperative relay communication in cognitive radio systems, in *2012 IEEE International Conference on Communications (ICC)* (2012), pp. 1550–1554

54. R. Meng, Y. Ye, N. gang Xie, Multi-objective optimization design methods based on game theory, in *2010 8th World Congress on Intelligent Control and Automation* (2010), pp. 2220–2227

55. H. Xu, B. Li, Efficient resource allocation with flexible channel cooperation in OFDMA cognitive radio networks, in *2010 Proceedings IEEE INFOCOM* (2010), pp. 1–9

56. H. Xu, B. Li, Resource allocation with flexible channel cooperation in cognitive radio networks. IEEE Trans. Mobile Comput. **12**(5), 957–970 (2013)

57. R. Xie, F. Yu, H. Ji, Spectrum sharing and resource allocation for energy-efficient heterogeneous cognitive radio networks with femtocells, in *2012 IEEE International Conference on Communications (ICC)* (2012), pp. 1661–1665

58. R. Xie, F. Yu, H. Ji, Y. Li, Energy-efficient resource allocation for heterogeneous cognitive radio networks with femtocells. IEEE Trans. Wirel. Commun. **11**(11), 3910–3920 (2012)

59. D. Nguyen, M. Krunz, Heterogeneous spectrum sharing with rate demands in cognitive MIMO networks, in *2013 IEEE Global Communications Conference (GLOBECOM)* (2013), pp. 3054–3059

60. M. Venkatesan, A. Kulkarni, Soft computing based learning for cognitive radio. Int. J. Recent Trends Eng. Technol. **10**(1), 12 (2014)

61. E. Shakshuki, M. Younas, A. Ahmed, G. Amel, S. Anis, M. Abdellatif, ANT 2012 and MobiWIS 2012 resource allocation for multi-user cognitive radio systems using multi-agent Q-learning. Procedia Comput. Sci. **10** 46–53 (2012). http://www.sciencedirect.com/science/article/pii/S1877050912003675

62. Y. Huang, J. Wang, H. Jiang, Modeling of learning inference and decision-making engine in cognitive radio, in *2010 Second International Conference on Networks Security, Wireless Communications and Trusted Computing* **2**, 258–261 (2010)

63. G.V. Lakhekar, R.G. Roy, A fuzzy neural approach for dynamic spectrum allocation in cognitive radio networks, in *2014 International Conference on Circuits, Power and Computing Technologies [ICCPCT-2014]* (2014), pp. 1455–1461

Chapter 6
Modelling and Analyses of Resource Allocation Optimisation in Cognitive Radio Networks

6.1 Need for Resource Allocation Models in Cognitive Radio Networks

Usually, new and/or emerging technologies follow some well-defined developmental processes or stages in their evolution, from the initial conception of the ideas to the final realisation of the usable product, service or application. After the initial *conceptualisation* of the idea, the developmental stages for new technologies usually involve some form of *computer modelling* of a prototype of the particular technology of interest. After computer models have been developed, this is most likely followed by the *simulation* or *emulation* of the models or prototypes. Thereafter, *experimentation* (using test-beds, for example) of the models are carried out.

Still, experimental designs are usually carried out alongside or followed by the *numerical analyses* of the models or prototypes (there are instances too where numerical analyses are carried out before experimentation or test-bed designs). By leveraging the initial results obtained through the experimentation and analyses, a lot more *fine-tuning* is carried out on the models or prototypes to obtain the best results possible. Only after sufficient fine-tuning has been achieved are the new technologies (products, applications, etc.) *rolled out*. Of course, there would be necessary reviews and *feedback*, which will usually lead to further improvements in the various *versions* of the developing technologies in the long run.

It is interesting to note that most emerging next-generation (xG) wireless communication technologies do follow the same developmental processes highlighted above in their evolution. The cognitive radio networks (CRN), being an emerging xG network itself, is also undergoing some of these evolutionary stages. Network modelling is a very significant aspect of technological evolution. As the CRN evolves, several models are being developed to study different aspects of it, particularly the aspect of resource optimisation for practical CRN realisation.

© The Author(s), under exclusive license to Springer Nature Switzerland AG 2022
B. TJ Maharaj, B. S. Awoyemi, *Developments in Cognitive Radio Networks*,
https://doi.org/10.1007/978-3-030-64653-0_6

Since there are already a plethora of works and materials that are currently available on modelling of resource allocation (RA) for the CRN, it is almost impossible to identify, study, analyse and discuss each one of them in details. Therefore, the approach that we use for our discussion on RA modelling for the CRN is to develop generic but very useful RA models, which we then study and analyse. To a large degree, the RA modelling for the CRN, as discussed in this chapter, can be used to interpret and study almost all RA models and designs for modern CRN applications.

Further, the RA models that are discussed in this chapter do have some important characteristics and/or advantages over most other RA models for the CRN. The first is that they are quite generic in their design. The RA models are generic in that they capture very succinctly the most essential elements or components of the modern CRN designs. Secondly, these RA models take into consideration the important aspect of heterogeneity for the CRN. This makes the RA models to be more accurate in their representation of the modern CRN. The third advantage of the RA models is that it is possible to analyse them in order to achieve both optimal and near-optimal solutions for the RA problems that are developed. Another important advantage is that the RA models discussed can be easily modified to accommodate any other (new) objective or constraint(s) without significantly changing the outlook, analyses and complexities of the RA problems or solutions for the CRN. Finally, the RA models that are developed are implementable in practical CRN applications.

6.2 System Modelling for Resource Allocation in Heterogeneous Cognitive Radio Networks

The system model shown in Fig. 6.1 is a generic model for heterogeneous CRN. The model is applicable for studying and analysing RA problems and solutions in modern CRN. The model is well-designed in that it incorporates the different kinds of heterogeneous considerations or classifications for the CRN. For example, the concept of *network heterogeneity* is well captured in the model. Network heterogeneity is achieved by separating the secondary network from the primary network, while each network is controlled by its own base station. Furthermore, network heterogeneity is achieved in the model by enabling each network to operate using different configurations of transmission power, modulation schemes, interference threshold, etc.

The concept of *channel heterogeneity* is also incorporated in the model in Fig. 6.1. Channel heterogeneity is achieved by the use of the orthogonal frequency division multiple access (OFDMA) technique for the CRN. This makes it possible for different secondary users (SUs) to use different parts of the spectrum band of the primary users (PUs) at the same time. Some newer variants of the OFDMA, and

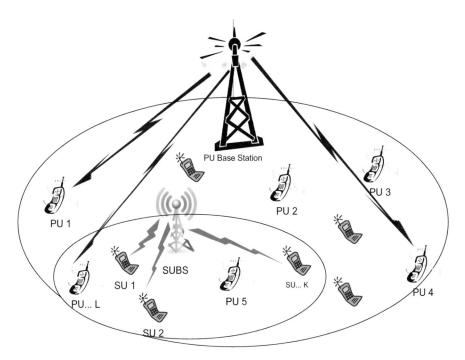

Fig. 6.1 A generic system model for heterogeneous CRN

even the non-orthogonal multiple access (NOMA) technique, are all useful medium access techniques to help realise channel heterogeneity for the CRN. Finally, the concept of *user heterogeneity* is incorporated in the model since all the SUs fall into classes or groups and each class is served based on some predetermined criteria.

The heterogeneous CRN model presented in Fig. 6.1 is a *centralised* model. However, by some slight modification, the model can be employed to study the distributed CRN representation as well. In the centralised CRN being considered, the SUBS has the responsibility of informing the SUs on the resources (modulation scheme, data rate, subchannels, transmission power, etc.) that have been allotted or assigned to them. However, in the distributed CRN, there is no central control. Each SU determines its resource usage by some ground rules that are applicable to all SUs in the network. More importantly, the *underlay*, *overlay* and *hybrid* architectural designs for the CRN, which are the most common descriptions of the CRN, are studied by slightly modifying the model to fit the various classifications. In the remaining parts of this section, we develop the generic system model to fit the underlay, overlay and hybrid CRN considerations.

6.2.1 Modelling Underlay Cognitive Radio Networks

The underlying ideas for the development of a generic RA model for the underlay CRN are presented in this subsection. In the underlay consideration, the SUs are free to transmit their data within the interference range of the PUs, but they must do so using very low power levels that they do not cause excessive interference to the PUs [1]. In the underlay model, we assume that there are N OFDMA subchannels within the coverage region of the secondary user base station (SUBS). Further, within the coverage range of the SUBS, there are K heterogeneous SUs and L similar PUs geographically dispersed in that region or space.

Since it is a centralised network, in the underlay, overlay and hybrid RA models studied, the SUBS is responsible for selecting the subchannels for each SU in the network. The SUBS then sends its decision to the SU on a distinct control channel. We assume seamless communication between the SUs and the SUBS over the control channel. We also assume a slow fading environment for the subchannels. The data rate c achieved by a subchannel depends on the modulation scheme through which that subchannel transmits its data. Four possible modulation schemes are considered in the model, namely, the binary phase shift keying (BPSK), the 4-quadrature amplitude modulation (4-QAM), the 16-QAM and the 64-QAM. The respective data rates for the four modulation schemes are $c = 1, 2, 4$ and 6 bits per OFDMA symbol. A rate weight w ($w > 0$) is associated with each category of SUs in the network.

Given that a particular value of the bit error rate (BER) ρ is to be realised, the BPSK modulation requires a minimum transmission power $P(c, \rho)$ given as [2]:

$$P(c, \rho) = N_\phi [c \times erfc^{-1}(2\rho)]^2$$
$$(c = 1).$$

(6.1)

Similarly, for the M-ary QAM, the minimum transmission power that is required is given as:

$$P(c, \rho) = \frac{2(2^c - 1)N_\phi}{3} \left(erfc^{-1} \left(\frac{c\rho\sqrt{2^c}}{2(\sqrt{2^c - 1})} \right) \right)^2$$

$(c = 2, 4$ or 6 for $4 - QAM, 16 - QAM$ and $64 - QAM,$ respectively),

(6.2)

where $erfc(x) = (\frac{1}{\sqrt{2\pi}}) \int_x^\infty e^{\frac{-t^2}{2}} dt$ is the complementary error function, $\pi = (22/7)$ and N_ϕ is the single-sided noise power spectral density. The value of N_ϕ is taken to be the same value for all the subchannels.

To maintain a particular ρ value, an increase in the number of bits on a subchannel will result in a non-linear increase in the amount of transmission power required to communicate on that subchannel. The subchannel power gain matrix between SUs and the SUBS is given as $\boldsymbol{H}^s \in R^{K \times N}$. Therefore, $H^s_{k,n}$ represents the power gain between the SUBS and the SU k on subchannel n. The minimum power $P_{k,n}(c_{k,n}, \rho)$ that SU k needs on subchannel n to transmit $c_{k,n}$ bits is calculated by dividing the power $P(c_{k,n}, \rho)$ of SU k on subchannel n by the channel gain $H^s_{k,n}$ between the SUBS and SU k on subchannel n. This is given as:

$$P_{k,n}(c_{k,n}, \rho) = \frac{P(c_{k,n}, \rho)}{H^s_{k,n}}. \tag{6.3}$$

We denote the power gain matrix between the PUs and the SUBS by $\boldsymbol{H}^P \in R^{L \times N}$. Therefore, the vector $\boldsymbol{H}^p_{l,n}$ represents the subchannel power gain between the SUBS and PU l on subchannel n.

The parameters so far defined and represented are the basic parameters for modelling the RA problems in the underlay CRN.

6.2.2   Modelling Overlay Cognitive Radio Networks

The underlying ideas for the development of a generic RA model for the overlay CRN are presented in this subsection. In the overlay model, the K heterogeneous SUs use the PUs' licensed spectrum in an opportunistic manner. There are L PUs licensed to use the spectrum. In this case, the entire spectrum space is divided into M subchannels. However, through periodic spectrum sensing, a subset N of the M subchannels is identified and selected to transmit data for the SUs.

Thus, the selected spectrum band N are the non-active frequency bands of the PUs during the particular time period of operation. This means that the SUs only use the subchannels that are in the vacant sub-bands of the PUs for their communication. Besides, all the other definitions and representations for the underlay RA model already discussed in the previous subsection (data rates, modulation schemes, transmission power, BER, weight, etc.) are applicable to the overlay RA model being discussed in this chapter. The main difference is that, in the overlay modelling, the possibilities of *miss detection* and *false alarms* are taken into consideration in the design and analysis. Miss detection brings about the possibility of co-channel interference to the PUs. False alarm usually results in the minimisation of the utilisation efficiency of the spectrum for the CRN.

Let the probability of miss detection on subchannel n be P^{md}_n and the probability of false alarm on subchannel n be P^f_n. Let $P_{1,n}$ be the probability that subchannel n is truly occupied by a PU and the CRN detects this correctly, then:

$$P_{1,n} = \frac{P^L_n(1 - P^{md}_n)}{P^L_n(1 - P^{md}_n) + P^f_n(1 - P^L_n)}, \tag{6.4}$$

where P_n^L is the priori probability that the sub-band of subchannel n is occupied by PUs. Again, let $P_{2,j}$ be the probability that subchannel j is truly occupied but the CRN has made a wrong judgement (that is, adjured it to be free). Then, an interference value of $I_{l,n}$ is experienced by PU l because an SU has accessed the subchannel n to transmit with unit transmission power. The interference $I_{l,n}$ is given as:

$$I_{l,n} = \sum_{j \in M_o^l} P_{1,j} I_{l,j}^n + \sum_{j \in M_v^l} P_{2,j} I_{l,j}^n, \tag{6.5}$$

where M_o^l is the set of subchannels that are sensed to be *occupied* in the sub-band of PU l, M_v^l is the set of subchannels that are sensed to be *vacant* in the sub-band of PU l and $I_{l,j}^n$ is the interference to subchannel j, which is in the sub-band of PU l, when an SU is transmitting on subchannel n with unit power.

6.2.3 Modelling Hybrid Cognitive Radio Networks

The underlying ideas for the development of a generic RA model for the hybrid CRN are presented in this subsection. In the hybrid model, the K heterogeneous SUs use the licensed spectrum of the PUs when the PUs are not available. Even when the PUs are around, the SUs are still allowed to use the PU sub-bands by simply reverting to the low power transmission (underlay) mode at such instances. There are L PUs licensed to use the spectrum. If N represents the number of available or inactive subchannels (i.e., the subchannels that are free because they are unoccupied by the PUs) and M represents the number of unavailable or active subchannels (i.e., the subchannels that are being used by the PUs), then the total available spectrum sub-bands in the hybrid case is actually the entire $(N + M)$ subchannels.

Again, in the hybrid CRN consideration, the SUs share both the active subchannels and the non-active subchannels with the PUs. For the hybrid model being considered, it is assumed that the activities of the PUs are stable during each time period or time frame. All the other definitions and representations for the underlay and overlay RA models discussed in the previous subsections (data rates, modulation schemes, transmission power, BER, weight, etc.) are applicable to and holds for the hybrid RA model developed.

6.3   Representing User Heterogeneity in the Resource Allocation Modelling for Cognitive Radio Networks

So far in our network modelling for the underlay, overlay and hybrid CRN considerations, the aspects of network heterogeneity and channel heterogeneity have been well represented in the RA models developed. Albeit, the aspect of user heterogeneity has not been captured in these models. This is because user heterogeneity is usually captured by defining different constraints for different user categories in the RA problem formulations. In this section, a general rule for capturing different categories of user heterogeneity in the RA modelling for the CRN is presented.

We assume that the K heterogeneous SUs in each design (underlay, overlay and hybrid CRN) are categorised into v different groups or classes of SUs. Any useful criterion can be used to classify these SUs. An example of a useful criterion for classifying SUs is the minimum data rate requirement for an acceptable quality of service (QoS). We number the different classes of SUs as $1, 2, 3, \ldots, v$, so that K_1 represents the number of SUs in class 1, K_2 represents the number of SUs in class 2 and so on. We attach weight w_i to the SUs in class $i \in v$. Therefore, w_1 is the weight attached to meeting the demands of the SUs in class 1, and w_v is the weight attached to meeting the demands of the SUs in v. We use throughput maximisation as our objective in the RA problem formulation. Therefore, the objective of the CRN design is to maximise the total data rate for all the SUs in all classes, which is the same as the throughput of the network. The objective function is then given as:

$$\max z = \sum_{n=1}^{N} \left(\sum_{k=1}^{K_1} w_1 c_{k,n} + \sum_{k=K_1+1}^{(K_1+K_2)} w_2 c_{k,n} \right.$$

$$+ \sum_{k=K_1+K_2+1}^{(K_1+K_2+K_3)} w_3 c_{k,n} + \ldots + \left. \sum_{k=K_1+\ldots+K_{v-1}+1}^{(K_1+K_2+\ldots+K_v)} w_v c_{k,n} \right);$$

$$c_{k,n} \in \{0, 1, 2, 4, 6\}. \tag{6.6}$$

To make the classification more practical, we assume that the K heterogeneous SUs are classified using their minimum data rate demand. If R_1 represents the minimum rate demand of the SUs in class 1, R_2 represents the minimum rate demand for the SUs in class 2 and so on, then R_v will be the minimum rate demand for the SUs in class v. We now write the minimum rate constraints for the different classes of SUs as follows:

$$\sum_{n=1}^{N} c_{k,n} \geq R_1; \ k = 1, 2, \cdots, K_1 \tag{6.7}$$

$$\sum_{n=1}^{N} c_{k,n} \geq R_2; \ k = K_1 + 1, K_1 + 2, \cdots, K_1 + K_2 \tag{6.8}$$

$$\sum_{n=1}^{N} c_{k,n} \geq R_3; \ k = K_1 + K_2 + 1, K_1 + K_2 + 2, \cdots, K_1 + K_2 + K_3 \tag{6.9}$$

$$\vdots$$

$$\sum_{n=1}^{N} c_{k,n} \geq R_v; \; k = (K_1 + K_2 + \ldots + K_{v-1} + 1),$$

$$(K_1 + K_2 + \ldots + K_{v-1} + 2), \cdots, (K_1 + K_2 + \ldots + K_v). \tag{6.10}$$

The equations given above are used to capture the different classes of users or user demands in any heterogeneous CRN consideration. While we have used the minimum data rate as a criterion for classifying the SUs, any other criterion could have been used to classify the SUs without significantly changing the format or shape of the representation.

6.4   General Formulation of Resource Allocation Problems in Cognitive Radio Networks

In this section, the various components of the RA modelling for the CRN discussed in the previous sections are finally combined to present generic RA formulations for the underlay, overlay and hybrid heterogeneous CRN. For ease of representation and analysis, it is assumed that the SUs are classified using their *minimum rate requirements*. Further, to keep the model as simple and straightforward as possible, we categorise the K heterogeneous SUs into just two classes. The class 1 SUs are represented by K_1 and they are the high-rate demand (HD) SUs. The class 2 SUs are represented by $(K - K_1)$ and they are the low-rate demand (LD) SUs. Each class has a minimum data rate demand for an effective QoS realisation.

We stress that we have only limited the classification of the heterogeneous SUs in the discussions and analyses presented in this chapter to two classes. The reason is simply to make the RA models quite manageable and easy to analyse and understand. However, it is very easy to extend the classes of SUs to three, four or any desired number of classes, using any classification criteria, following the general formulation approach presented in the previous section. Increasing the number of classes of SUs will not significantly change the solutions or results that are obtained for the CRN.

Using the definitions and representations that have so far been provided in the previous sections of this chapter, the formulation of the generic **underlay** RA optimisation problem for heterogeneous CRN, considering the minimum rate demands of the different classes of SUs as the criterion for categorising the SUs, is presented as follows:

$$\max z = \sum_{n=1}^{N} \left(\sum_{k=1}^{K_1} w_1 c_{k,n} + \sum_{k=K_1+1}^{K} w_2 c_{k,n} \right); \tag{6.11}$$

$$c_{k,n} \in \{0, 1, 2, 4, 6\}$$

subject to

$$\sum_{n=1}^{N} c_{k,n} \geq R_I; \quad k = 1, 2, \cdots, K_1 \tag{6.12}$$

$$\sum_{n=1}^{N} c_{k,n} \geq R_{II}; \quad k = K_1 + 1, K_1 + 2, \cdots, K \tag{6.13}$$

$$\sum_{n=1}^{N} \sum_{k=1}^{K} P_{k,n} \leq P_{\max} \tag{6.14}$$

$$\sum_{n=1}^{N} \Phi_n H_{l,n}^{p} \leq \varepsilon_l; \quad l = 1, 2, \ldots, L \tag{6.15}$$

$$c_{k,n} = 0 \; if \; c_{k',n} \neq 0, \quad \forall k' \neq k; \quad k = 1, 2, \ldots, K \tag{6.16}$$

where R_I is the minimum data rate demand of SU k in class 1 and R_{II} is the minimum data rate demand of SU k in class 2, w_1 is the weight attached to the class 1 SUs and w_2 is the weight attached to the class 2 SUs, $\Phi_n = \Sigma_{k=1}^{K} P_{k,n}$ is the total power on subchannel n, $P_{k,n}$ is the transmission power of SU k on subchannel n, $H_{l,n}^{p}$ is the magnitude of the interference channel gain between the SUBS and PU l on subchannel n, ε_l is the threshold interference power to PU l from all the SUs in the network and $P_{\max}$ is the SUBS' maximum transmission power.

Equation (6.11) captures the objective function for the RA problem in underlay heterogeneous CRN. In this case, the objective function is indicative of the sum throughput or total weighted data rate that is realised by all the SUs in the network. The constraints in Eqs. (6.12) and (6.13) are the minimum data rate constraint. These constraints ensure that the respective minimum data rate requirements for the two classes of SUs are satisfied. The constraint in Eq. (6.14) ensures that the total transmission power of all the SUs do not exceed the maximum transmission power of the SUBS. The constraint in Eq. (6.15) is to ensure that the amount of interference that reaches each PU when the SUs are transmitting on the PU's channel do not exceed the set threshold interference value. The constraint in Eq. (6.16) is the mutually exclusive constraint. The constraint ensures that only one SU is assigned to each subchannel. Therefore, once we have allocated subchannel n to the SU $k' \neq k$, the data rate for subchannel n must be 0 for any other user k.

In the case of the **overlay** model, the formulation of the generic RA optimisation problem for heterogeneous CRN, using the minimum rate demands of the different classes of SUs to categorise the SUs, is the same formulation for the underlay CRN, as presented in Eqs. (6.11)–(6.16), EXCEPT for Eq. (6.15) which captures the interference constraint to the PUs. For the overlay RA model, Eq. (6.15) becomes:

$$\sum_{n=1}^{N} \Phi_n H_{l,n}^{p} I_{l,n} \leq \varepsilon_l; \ l = 1, 2, \ldots, L, \tag{6.17}$$

in which case, $I_{l,n}$ represents the possible interference that is caused to the PUs as a result of the problems of miss detection and false alarm (already given in Eq. 6.5). All other parts of the problem formulation in Eqs. (6.11)–(6.16) are unchanged for the overlay CRN.

For the **hybrid** CRN consideration, the formulation of the generic RA optimisation problem for heterogeneous CRN, using the minimum rate demands of the different classes of SUs to categorise the SUs, is slightly modified to reflect that all available subchannels (occupied or unoccupied) are used by the SUs. In this case, the general formulation given in Eqs. (6.11)–(6.16) becomes:

$$\max z = \sum_{n=1}^{(N+M)} \left(\sum_{k=1}^{K_1} w_1 c_{k,n} + \sum_{k=K_1+1}^{K} w_2 c_{k,n} \right); \tag{6.18}$$

$$c_{k,n} \in \{0, 1, 2, 4, 6\}$$

subject to

$$\sum_{n=1}^{(N+M)} c_{k,n} \geq R_I; \quad k = 1, 2, \cdots, K_1 \tag{6.19}$$

$$\sum_{n=1}^{(N+M)} c_{k,n} \geq R_{II}; \quad k = K_1 + 1, K_1 + 2, \cdots, K \tag{6.20}$$

$$\sum_{n=1}^{(N+M)} \sum_{k=1}^{K} P_{k,n} \leq P_{\max} \tag{6.21}$$

$$\sum_{n \in N} \Phi_n H_{l,n}^{p} + \sum_{n \in M} \Phi_n H_{l,n}^{p} I_{l,n} \leq \varepsilon_l; \quad l = 1, 2, \ldots, L \tag{6.22}$$

$$c_{k,n} = 0 \ if \ c_{k',n} \neq 0, \ \forall k' \neq k; \quad k = 1, 2, \ldots, K. \tag{6.23}$$

6.5   General Problem Formulation While Employing Other Heterogeneous User Classifications

To further broaden the scope of the generic RA problem formulations for the underlay, overlay and hybrid CRN considerations discussed in the previous section, we explore how the developed RA problems can be made to accommodate other possible heterogeneous classifications of the SUs. While the formulation in the previous section employed the minimum rate requirements to classify the heterogeneous SUs, that criterion alone is not the only criterion for the possible classification of heterogeneous SUs in the CRN. There are other criteria for classifying the SUs in the CRN which must also be put into consideration for the RA formulation to be indeed generic.

Again, since the RA models discussed in this chapter are meant to be generic, the general formulation of the RA problem should not be limited to cover heterogeneous user classification based on the minimum rate requirements alone. The generic RA problem formulation should be such that it can be easily extended to cater for other user categorisations as well. In other words, if the heterogeneous classification of the SUs were based on any other criterion, it should be possible to easily modify the generic RA problems, as discussed in the previous section, to now address such new classes of SUs. In this section, we seek to establish how the generic RA formulation that is developed in the previous section can be easily modified to capture other possible user classifications for wider applicability.

As previously mentioned, apart from the minimum rate requirements, there are other criteria for classifying user heterogeneity that have been studied in some of the RA models developed for heterogeneous CRN. For example, in the work in [3], a number of criteria for classifying heterogeneous users in the CRN were considered. Some of the most common criteria that have been employed for classifying heterogeneous SUs are the *minimum rate requirements* of the SUs [1, 4], the level of *priority or sensitivity* of the SUs [5, 6], the *delay tolerance profile* of the SUs [7, 8], among others.

The RA problems developed in the previous section (Eqs. (6.11)–(6.17)) can indeed be easily modified to accommodate almost all other criteria for classifying the SUs. In other words, the RA problems developed in the previous section are indeed generic in that, almost all the criteria for classifying user heterogeneity can be incorporated into the general RA model developed for the underlay, overlay and hybrid CRN. We provide an example of how this can be achieved. We use the instance where the SUs are classified based on their *priority* levels to achieve the above-mentioned claim. We consider only two categories of users still, for ease of representation and analysis. In the instance of priority classification, the SUs can be classified as either 'high priority (HP)' SUs or 'best effort service (BE)' SUs.

Using the above priority classification, the class 1 SUs, which are the HP SUs, will have the higher priority and their demands will always be met first. The class 2 SUs, which are the BE SUs, will share what is left of the resources among themselves, using a proportional fairness constraint. The total number of heterogeneous SUs K is now divided into the two classes of SUs. Let K_1 represent the HP SUs and K_2 represent the BE SUs. The sets of the two classes of SUs are denoted as κ_A and κ_B, respectively.

Furthermore, we use R_k to represent the minimum data rate demand of SU k in κ_A and γ_k to represent the predefined value of the normalised proportional fairness factor for each SU in κ_B. Also, R_i represents the data rate for the element i in κ_B, w_1 represents the weight of SU k in κ_A and w_2 represents the weight of SU k in κ_B. In comparison, all other parts of the RA problem formulation in Eqs. (6.11)–(6.17) are unchanged EXCEPT for the constraints in Eqs. (6.12) and (6.13). The constraints in Eqs. (6.12) and (6.13), which captured the minimum data rate requirement in the previous classification, are now modified to capture the new priority classification for the heterogeneous SUs as follows:

$$\sum_{n=1}^{N} c_{k,n} \geq R_k; \; \forall k \in \kappa_A \tag{6.24}$$

$$\frac{R_k}{\sum_{i \in \kappa_B} R_i} = \gamma_k; \; \forall k \in \kappa_B, \tag{6.25}$$

in which case, the new Eq. (6.18) now satisfies the HP SUs while the new Eq. (6.19) now satisfies the BE SUs.

In like manner, all other heterogeneous user classifications can be easily accommodated by slightly adjusting the generic RA problem formulation of Eqs. (6.11)–(6.23). The work in [3] gives some practical examples of other heterogeneous classifications and how the generic RA formulation can be easily modified to accommodate them without significantly affecting the problem structure, analyses and results of the RA problems for the CRN. In light of the discussions so far presented, we may generalise that almost all the constraints or categories/classes of heterogeneity in the RA problems for the CRN can be accommodated by slightly modifying the RA problems developed in Eqs. (6.11)–(6.17). Therefore, *we make the case that the RA problems developed and discussed in this chapter are indeed generic RA optimisation problems for the underlay, overlay and hybrid heterogeneous CRN considerations.*

6.6 Relating Other Resource Allocation Problem Formulations to the General Formulation

Although we have attempted to present a broadly generic RA problem formulation for the underlay, overlay and hybrid CRN, it is still necessary to point out that the generic RA problem formulations developed and discussed in this chapter are not the only possible RA formulations for the CRN. Indeed, there are several other RA formulations for the CRN that have been presented and/or analysed in the literature. Obviously, a good number of the other RA formulations would likely have some kind of variation to the generic RA formulations for the underlay, overlay and hybrid heterogeneous CRN developed in this Chapter.

In reality, since individual presentations and goals are different, RA problem formulations for the CRN are usually diverse from one research work to another such that it may be difficult to find the RA problem formulations of two different works to be exactly alike in all respects, assumptions, analyses, etc. Therefore, to further establish the generic nature of the RA problem formulation developed in this chapter, in this section, we attempt to relate the generic problem formulations in this chapter to a few other RA problem formulations that have been developed by other researchers in the field.

6.6.1   Other Underlay Formulations in Relation to the General Formulation

There is already a sizeable volume of RA problem formulations for the underlay CRN that have been developed in various research works in the literature. These RA problem formulations may not be exactly like or completely similar to the generic RA problem formulation for heterogeneous underlay CRN presented in the previous sections of this chapter. However, a good number of the RA problem formulations may be easily related to the generic RA problem formulation developed and discussed in this chapter. We examine some works on RA for underlay CRN and establish how these models relate to the generic RA model for the underlay CRN presented in this chapter.

The authors in [9] developed and analysed an RA model for centralised, underlay CRN. In the model, there were K SUs, L PUs and N subchannels. The power gain between the SUBS and PU l at subchannel n was represented by $H_{l,n}^{p}$, the maximum permissible interference to PUs was represented by Υ_{l}, the total power allocated to subchannel n was represented by ϕ_{n}, the maximum available power at the SUBS was represented by $P_{\max}$ and the minimum required data rate for each SU k was represented by R_{k}. Furthermore, the transmission power needed at the SUBS to transmit $c_{k,n}$ bits to SU k on subchannel n with a BER threshold ρ was represented by $P_{k,n}(c_{k,n}, \rho)$, the set of all possible values for $c_{k,n}$ was represented by $D = 0, d_{1}, \ldots, d_{C} = 0, 1, 2, 4, 6$, where C stands for the number of possible modulation schemes ($c_{k,n} = 0$ meant that SU k did not use subchannel n for data transmission, $c_{k,n} = 1$ meant that SU k used subchannel n for BPSK transmission and so on). The objective was to maximise the sum throughput for the network. The optimisation problem in [9] was formulated as:

$$\max_{c_{k,n} \in D} \sum_{n=1}^{N} \sum_{k=1}^{K} c_{k,n} \tag{6.26}$$

subject to

$$\sum_{n=1}^{N} \phi_{n} H_{l,n}^{p} \leq \Upsilon_{l}, \quad l = 1, 2, \ldots, L, \tag{6.27}$$

$$\sum_{n=1}^{N} c_{k,n} \geq R_{k}, \quad k = 1, 2, \ldots, K, \tag{6.28}$$

$$\sum_{n=1}^{N}\sum_{k=1}^{K} P_{k,n}(c_{k,n}, \rho) \le P_{\max}, \tag{6.29}$$

$$c_{k,n} = 0, \ if \, c_{k',n} \ne 0, \ \forall k \ne k', \ k = 1, 2, \ldots, K. \tag{6.30}$$

The constraint in Eq. (6.27) ensured that the interference leakage to PU l was always below the threshold value Υ_l, the constraint in Eq. (6.28) ensured that each SU achieved the minimum required data rate R_k, the constraint in Eq. (6.29) limited the total transmission power below the available power $P_{\max}$ at the SUBS, the constraint in Eq. (6.30) ensured mutual exclusive SU allocation to each subchannel.

In comparison with the generic RA problem for underlay CRN developed in Eqs. (6.11)–(6.16), the RA problem in [9] is a very fair representation of the underlay CRN. Most of the important details for the underlay CRN are well-captured, except for the aspects of heterogeneous users and the effects of weight. The general RA formulation is thus a fuller and more comprehensive representation of the RA problems for underlay CRN than the problem presented in [9].

Another good example of the RA problem formulation for underlay CRN is found in [10]. In the model, the authors developed a multi-carrier CRN composed of a primary cell and N secondary cells. The secondary cells could use either the channel underlay or the channel interweave (hybrid) approach in accessing the PU subchannels. Index 1 referred to the primary network while indexes i with $i \in \{2, \ldots, N + 1\}$ referred to the secondary networks. The primary network occupied a licensed bandwidth B with L adjacent subchannels using the OFDMA technique. The SUBS sought to access the same bandwidth B for the SUs. The SUs cooperated for multiple access, allowing only one SU to transmit its data in each subchannel. The SU that was selected to use subchannel $k \in 1, \ldots, L$ was denoted by $i[k]$. The rate on link $i \in \{2, \ldots, N + 1\}$ at iteration n was denoted by R_i^k, the set of subchannels that were allocated to SU i was denoted by S_i, the function of the primary power at iteration $n1$ that indicated the channel gains was denoted by $b_{i,(n-1)}^k$, the different rate values which were determined by the different secondary network cases employed in the system were represented by a_i^k and c_i^k, the channel gain was denoted by $h_{1,i}^k$ and the threshold interference for subchannel k was denoted by I_{th}^k. The objective was to maximise the sum rate on the primary link and on the secondary links. Subchannel allocation was first carried out to select one secondary transmitter per subchannel. Thereafter, an iterative power allocation algorithm was used to allocate power to each SU. The optimisation problem for each SU $i \in \{2, \ldots, N + 1\}$ at iteration n in [10] was formulated as:

$$\max_{P_{i,n}} \frac{B}{L} \sum_{k \in S_i} \log_2(1 + b_{i,(n-1)}^k P_{i,n}^k) \tag{6.31}$$

subject to

$$a_i^k P_i^k \geq c_i^k \;\; \forall k \in S_i \tag{6.32}$$

$$|h_{1,i}^k|^2 P_i^k \leq I_{th}^k \;\; \forall k \in S_i \;\; if \, P_{1,(n-1)}^k > 0 \tag{6.33}$$

$$P_i^k \geq 0 \;\; \forall k \in S_i \tag{6.34}$$

$$\sum_{k \in S_i} P_i^k \leq P_{i,\max} \;\; \forall k \in S_i. \tag{6.35}$$

The constraint in Eq. (6.32) indicated the achievable rate on each subchannel, the constraint in Eq. (6.33) expressed the interference limitation on the primary receiver, while the last two constraints were the power constraints for the secondary network.

In comparison with the generic RA problem for underlay CRN developed in Eqs. (6.11)–(6.16), the RA problem in [10] is fairly comprehensive, even though its network classifications do not properly represent heterogeneous users in typical CRN. Furthermore, its problem formulation is slightly cumbersome, making it more complex for proper analysis than in the generic underlay RA problem formulation developed for heterogeneous CRN.

6.6.2   Other Overlay Formulations in Relation to the General Formulation

Just as in the underlay consideration, there is already a sizeable volume of RA problem formulations for the overlay CRN that have been developed in various research works in the literature. These RA problem formulations may not be exactly like or completely similar to the generic RA problem formulation for heterogeneous overlay CRN presented in the previous sections of this chapter. However, a good number of the RA problem formulations may be easily related to the generic RA problem formulation developed and discussed in this chapter. We examine some works on RA for overlay CRN and establish how these models relate to the generic RA model for the overlay CRN presented in this chapter.

The authors in [11] developed an overlay CRN model that is, in fact, very similar to the generic model described in this chapter. In the model, the SUs used the licensed spectrum of the PUs in an opportunistic manner with the aid of an access point (the equivalence of the SUBS). There were L PUs, K heterogeneous SUs (divided in K_0 non-real-time (NRT) SUs with each SU allotted resources by a rate proportionality factor γ_k, and $K - K_0$ real-time (RT) SUs with fixed rate requirements R_k^{req}) and N subchannels detected to be free by the SUs and over which they were permitted to transmit their data. The rate of SU k was represented by R_k, the interference introduced to PU l by the access of a SU on subchannel n with unit transmission power was denoted by $I_{n,l}$, the interference threshold for PU l was denoted by I_l^{th}, the power allocated to SU k on subchannel n was denoted by $p_{k,n}$, the signal-to-noise ratio (SNR) of subchannel n being used by SU k with unit power was represented by $H_{k,n}$, the transmission power limit of the SUBS was denoted by P_T and $\rho_{k,n}$ denoted the subchannel allocation index which was either 1 (indicating that the subchannel n has been occupied by SU k) or 0 (indicating otherwise). The objective of the RA problem for the overlay CRN developed in [11] was to maximise the downlink sum capacity of the SUBS, with guarantee on the rate demands of all SUs, under power limitation and interference constraints. The RA optimisation problem was formulated as:

$$\max_{\rho_{k,n}, p_{k,n}} \sum_{k=1}^{K} \sum_{n=1}^{N} \rho_{k,n} \log(1 + pk, n H_{k,n}), \tag{6.36}$$

subject to

$$p_{k,n} \geq 0, \quad \forall n \in N, \; \forall k, \tag{6.37}$$

$$\sum_{k=1}^{K} \sum_{n=1}^{N} \rho_{k,n} pk, n \leq P_T, \tag{6.38}$$

$$\sum_{k=1}^{K} \sum_{n=1}^{N} \rho_{k,n} pk, n I_{n,l} \leq I_l^{th}, \quad l = 1, 2, \ldots, L, \tag{6.39}$$

$$R_1 : R_2 : \ldots : R_{K_0} = \gamma_1 : \gamma_2 : \ldots : \gamma_{K_0}, \tag{6.40}$$

$$R_k = R_k^{req}, \quad k = K_0 + 1, \ldots, K, \tag{6.41}$$

$$\rho_{k,n} \in \{0, 1\}, \quad \forall n \in N, \; \forall k, \tag{6.42}$$

$$\sum_{k=1}^{K} \rho_{k,n} = 1, \quad \forall n \in N. \tag{6.43}$$

The constraint in Eq. (6.39) indicated that the permissible interference limit to PU l must not exceed its threshold value I_l^{th}. The constraint in Eq. (6.40) was the constraint that indicated the proportional rate of the NRT SUs. The constraint in Eq. (6.41) was the constraint that indicated the fixed rate requirements of the RT SUs. The constraints in Eqs. (6.42) and (6.43) were to ensure that each subchannel was not shared by more than one SU.

In comparison with the generic RA problem for overlay CRN developed in Eqs. (6.11)–(6.17), except for the weight implications for the different categories of SUs, the RA problem for overlay CRN developed in [11] is a very close substitute of the generic problem. All the important aspects of the overlay CRN captured in the generic formulation are equally represented in the problem formulation in [11]. The formulation in [11] is therefore a very detailed and a well-thought-out RA problem for overlay CRN.

Another good example of the RA problem for overlay CRN that can be compared with the general RA problem for the overlay CRN is found in [12]. In the model, there were G (index by g) multiple CRN systems (secondary networks) coexisting with a primary network. There were K heterogeneous SUs in the multiple CRN system, divided into single-network SUs (these SUs chose the best wireless networks that was available to access and send data through a single network at a time) and multi-homing SUs (these SUs accessed multiple networks simultaneously with multiple access technologies). There were N subchannels shared for the entire network. Through sensing, the subchannels in CRN g were divided into the subchannels that were available, N_g^a and the subchannels that were unavailable, N_g^u. The interference introduced to the PU in CRN g by the SU k over the subchannel n was denoted by $I_{k,g}^n$. The value of $I_{k,g}^n$ was below the permissible interference limit I_g^{th} of the PU in each CRN. The transmission power of SU k on subchannel n in CRN g was denoted by $p_{k,g}^n$, the total transmission power of SU k was denoted by $\bar{p}_k$, the rate capacity of SU k on subchannel n in CRN g was denoted by $R_{k,g}^n$, the minimum capacity requirement for each SU to guarantee the QoS demand was denoted by R_{min}, ψ_k was an indicator to show the type of SU that k was ($\psi_k = 1$ if SU k was a single network SU, $\psi_k = G$ if SU k was a multi-homing SU) and $\rho_{k,g}^n$ was the channel allocation indicator ($\rho_{k,g}^n = 1$ meant that subchannel n had been allocated to SU k in CRN g and $\rho_{k,g}^n = 0$ meant otherwise). The objective of the problem formulation was to maximise the total capacity of the multiple CRN system. The RA problem in [12] was formulated as:

$$\max_{\rho_{k,g}^n, p_{k,g}^n} \sum_{k=1}^{K} \sum_{g=1}^{G} \sum_{n \in N_g^a} R_{k,g}^n \qquad (6.44)$$

subject to

$$\sum_{k=1}^{K} \rho_{k,g}^n \leq 1, \quad \forall g, n, \qquad (6.45)$$

$$\rho_{k,g}^n + \rho_{k,g'}^{n'} \leq \psi_k, \quad \forall k, g \neq g', n, n', \qquad (6.46)$$

$$p_{k,g}^n - \bar{p}_k \rho_{k,g}^n \leq 0 \quad \forall k, gn, \qquad (6.47)$$

$$\sum_{g=1}^{G} \sum_{n \in N_g^a} p_{k,g}^n \leq \bar{p}_k \quad \forall k, \qquad (6.48)$$

$$\sum_{k=1}^{K} \sum_{n \in N_g^a} p_{k,g}^n I_{k,g}^n \leq I_g^{th} \quad \forall g, \qquad (6.49)$$

$$R_k \geq R_{min} \quad \forall k, \qquad (6.50)$$

$$p_{k,g}^n \geq 0 \quad \forall k, g, n, \qquad (6.51)$$

$$\rho_{k,g}^{n} \in \{0, 1\} \quad \forall k, g, n. \tag{6.52}$$

In comparison with the generic overlay RA problem developed in Eqs. (6.11)–(6.17), the overlay RA problem in [12] is much more cumbersome and quite intractable. There are so many extraneous parameters that render the problem unnecessarily complex and almost-certainly impracticable for real-life CRN applications. The general RA problem formulation for overlay CRN is thus a much better representation of the problem than the one developed in [12].

6.6.3 Other Hybrid Formulations in Relation to the General Formulation

Unlike the cases for underlay and overlay, the volume of work on RA problem formulations for the hybrid CRN is not very much. Still, a few works are cited and compared to the generic hybrid RA problem formulation developed in this chapter. The few RA problem formulations for hybrid CRN by other researchers may not be exactly like or completely similar to the generic RA problem formulation for heterogeneous hybrid CRN presented in the previous sections of this chapter. However, the few works on RA problem formulations may be easily related to the generic RA problem formulation developed and discussed in this chapter. We examine some works on RA for hybrid CRN and establish how these models relate to the generic RA model for the hybrid CRN presented in this chapter.

The authors in [13] developed an RA model for mobile SUs in a hybrid CRN environment to enhance spectrum efficiency. In the model, the activities of the PUs were assumed to be unchanged in each time frame. The SUs in the CRN were assumed to be mobile, with individual speed and direction for each SU in the CRN. The RA problem was defined to be a maximum throughput and fair access problem, where each SU sought fair access opportunities in such a way that the maximum expected throughput would be achieved for the entire CRN.

In the model developed in [13], the hybrid design that was employed was quite an interesting one. In the model, when the PU was unavailable, the SUs used the licensed subchannels of the PU at the highest power possible. When the PU was available, a power control mechanism was used to regulate the interference on the subchannels of the PU. In cases when the SUs were in very close interference range to the PU, the subchannels of the PU were completely unavailable to the SUs due to the uncontrollable interference they would cause. In essence, only the SUs at a comfortable distance to the PUs could access the subchannels at those times. Therefore, even though the hybrid sharing model was employed [13], there was still no guarantee that all the subchannels would always be available to all SUs at all the times.

For the analysis, the hybrid RA problem in [13] was developed as a subchannel assignment and power allocation problem. The objective was to maximise the expected network throughput and to guarantee access fairness for all the SUs. Since the PUs' activities were considered to be stable during each time period T, network throughput was optimised for only that period. The hybrid RA problem in [13] was formulated as:

$$\max z = \sum_{t=0}^{T} \sum_{n=1}^{N} \sum_{k=1}^{K} a_{k,n}(t) R_{k,n}(t) \tag{6.53}$$

subject to

$$\sum_{n=1}^{N} a_{k,n}(t) \geq 1, \tag{6.54}$$

$$\sum_{k=1}^{K} a_{k,n}(t) \leq 1; \ a_{k,n}(t) = \{0, 1\} \tag{6.55}$$

$$P_n(t) \leq P_n^{\mathrm{lim}}; \ \sum_{n=1}^{N} P_n(t) \leq P_{\max}, \tag{6.56}$$

where $a_{k,n}(t)$ represented the instantaneous subchannel allocation result, which was 1 when subchannel n was assigned to SU k and 0 otherwise, P_n^{lim} was the power limit on subchannel n and $P_{\max}$ was the maximum transmission power of the SUBS. The constraint in Eq. (6.54) guaranteed that there was at least one subchannel being assigned to each SU. The two constraints in Eq. (6.55) ensured that only one connection was permitted on each subchannel. The two constraints in Eq. (6.56) represented the power constraints. The optimisation problem in Eqs. (6.53)–(6.56) was time-varying and was said to be very difficult to solve directly. The authors solved the problem suboptimally by dividing it into a number of sub-problems, which were then solved per time using the approach of graph theory.

In comparison with the generic hybrid RA problem developed in Eqs. (6.18)–(6.23), the hybrid RA problem in [13] completely left out the important aspect of interference to PUs, making it incomplete and insufficient to adequately analyse hybrid RA for the CRN. All other parts of the RA problem in [13] are well captured in the generic RA problem for hybrid CRN. The generic RA problem is thus a more comprehensive and/or a more inclusive version of the RA problem developed in [13].

Another example of the RA problem for hybrid CRN that can be compared with the generic RA problem for hybrid CRN developed in this chapter is found in [14]. In the model, the SUs in the overlay region used the overlay method to access the spectrum, whereas the SUs in the hybrid region used the underlay or sensing-free method to access the spectrum. The OFDMA technique was used for the primary network and the secondary network. The PUs' spectrum band was divided into N equal-sized subchannels, while each subchannel experienced flat fading.

The authors assumed perfect sensing and did not consider the case of imperfect sensing. The network was centralised and not distributed. The objective was that of minimising the total power being consumed by the network. This was subject to meeting a minimum data rate demand. The RA problem for hybrid CRN in [14] was formulated as:

$$\min_{P_{i,k}\rho_{i,k}\forall_{i,k}} \sum_{k=1}^{K} \sum_{i\in N\cup M} \rho_{i,k} P_{i,k} \qquad (6.57)$$

subject to

$$R_k = \sum_{i\in N} \rho_{i,k} C\left(\frac{P_{i,k} h_{i,k}^{SS}}{\sigma^2}\right) + \alpha^{(k)} \sum_{i\in M} \rho_{i,k} C\left(\frac{P_{i,k} h_{i,k}^{SS}}{\sigma^2 + P_p h_i^{PS}}\right) \geq R^{\min} \quad \forall k, \qquad (6.58)$$

$$\sum_{i\in N\cup M} \rho_{i,k} P_{i,k} \leq P_k^{\max} \quad \forall k, \qquad (6.59)$$

$$\alpha^{(k)} \rho_{i,k} P_{i,k} L_{i,k}^{SP} \leq I_i^{\max} \quad \forall i \in M \ \forall k, \qquad (6.60)$$

$$\sum_{k=1}^{K} \rho_{i,k} \leq 1, \rho_{i,k} \in \{0, 1\} \quad \forall k, i \qquad (6.61)$$

where N represented the set of subchannels that were detected to be free or unoccupied, M represented the set of subchannels that were detected to be occupied or unavailable, $P_{i,k}$ denoted the transmission power allocated on subchannel i for SU k, P_p represented the transmission power of the PUBS, $R^{\min}$ denoted the minimum rate requirement of the SUs, $P_k^{\max}$ represented the power budget of SU k, $I_i^{\max}$ denoted the QoS threshold of subchannel i for the primary network, $h_{i,k}^{SS}$ represented the instantaneous channel gain on subchannel i from SU k to the SUBS, h_i^{PS} denoted the instantaneous channel gain on subchannel i from the PUBS to the SUBS, $L_{i,k}^{SP}$ represented the average channel gain on subchannel i from SU k to the worst-case PU, σ^2 denoted the noise power of each subchannel at the SUBS, $\alpha^{(k)}$ represented the spectrum sharing indicator of SU k ($\alpha^{(k)} = 0$ for SUs in the overlay region and $\alpha^{(k)} = 1$, otherwise) and $\rho_{i,k}$ represented the channel allocation indicator ($\rho_{i,k} = 1$ meant that subchannel i had been allocated to SU k and $\rho_{i,k} = 0$ meant otherwise).

In comparison with the general hybrid RA problem for the CRN presented in Eqs. (6.18)–(6.23), the RA problem in [14] is fairly comprehensive and detailed. The general RA problem maximises the overall network throughput (or the total data rate for the network), subject to a constraint on the maximum transmission power. On the other hand, the RA problem in [14] minimised the overall power consumption for the network, subject to a constraint on the data rate requirement. The interchange in the objectives for the two RA problems is completely in order.

Both power consumption and throughput are equally essential in the network design, therefore optimising either one (subject to the limiting effects of the other one) is perfectly alright. Even though the RA problem in [14] is well-developed and fairly comprehensive, it again omitted the aspect of interference to PUs, which makes it inadequate and incomplete. Thus, the general RA problem for hybrid CRN studied in this chapter is a more comprehensive RA representation for the CRN than the RA problem developed in [14].

6.7   Exploring Practical Solutions for the Resource Allocation Problems in Cognitive Radio Networks

The generic RA problem formulations for heterogeneous underlay, overlay and hybrid CRN, as developed and presented in the previous sections, are non-linear and non-deterministic polynomial-time hard (NP-hard) problems, particularly because of the power constraints in those formulations. Similarly, most other RA problems and problem formulations for the underlay, overlay and hybrid CRN are almost always non-linear and NP-hard problems. In an earlier chapter of this book, the non-linearity of the RA problems in the CRN was established, and also, appropriate optimisation tools to help solve such RA problems were discussed. In this section, we explore and employ one of the most-promising optimisation tools to solve the generic RA problems in the CRN, as formulated in this chapter.

   In almost all certainty, practicable solutions to and analyses of the RA problems in heterogeneous CRN would employ some of the well-developed optimisation solution tools and methods for solving RA problems in the CRN, as already well-documented in an earlier chapter of this book. In this section, therefore, we investigate and present a solution tool that combines the methods of *studying the problem structure* and *problem reformulation* in transforming the non-linear, NP-hard optimisation problems into the more-solvable integer linear programming (ILP) problems. The resulting ILP problems are easily solved through *classical optimisation*. The important advantage of the solutions provided through the combined methods explored in this section is that the solutions are optimal and are fairly practicable, especially if the heterogeneous CRN is not too large.

6.7.1   Studying the Structure of the Problems

To help solve the complex non-linear NP-hard RA problems for heterogeneous underlay, overlay and/or hybrid CRN, such as the ones presented in this chapter, a good practise is to examine the structure of the problem for any clues that may be explored in arriving at good solutions. In the case of the generic RA problems for the underlay, overlay and hybrid CRN discussed in this chapter, a careful consideration of the structure of the problem gives useful clues on how to solve them. By a careful observation of the structure of the problems, two important points are noted. These two striking points on the structure of the RA problems can be employed to achieve the ILP reformulation of the initial RA problems. These two important aspects of the structure of the RA problem are discussed.

The first observation on the structure of the RA problems for heterogeneous CRN is the fact that *only integer bits are allocated to the various subchannels in the network*. This is an important observation. It simply means that no fraction of bits can be assigned to any user at any given time of the network operation. In other words, all bit allocations to all the SUs in the CRN are simply in whole numbers or zero. The RA problem formulations make it impossible to allocate half bits or some other fractions of bits to any SU at any time of the network cycle.

The second important observation on the structure of the RA problems developed for heterogeneous CRN is that *the allocation of the network subchannels is an 'either' 'or' (binary) decision*. The subchannels are either allocated integer bit(s) to transmit data or they are not assigned bit(s) to transmit data. In more details, the decision on whether or not to give data rates to a subchannel to transmit data is made by considering the channel interference on that subchannel. If a subchannel has a channel interference gain to the PUs that is within some acceptable limit permitted by the PUs, that subchannel is allocated some bits to transmit data. However, if the channel interference gain to the PUs is high for a particular subchannel, such a subchannel will not be assigned any bit to transmit data, or at best, it will only be allocated very minimal bits to transmit data. This decision helps to ensure that the interference to the PUs is always minimal and within the acceptable limit to the PUs.

6.7.2 Problem Reformulation

The two important observations on the structure of the RA problems for heterogeneous CRN discussed in the previous subsection are exploited to achieve a linear reformulation of the initial RA problems developed in the previous sections of this chapter. The linear reformation process carried out in this subsection is applicable to the underlay, overlay and hybrid RA problems. The ILP reformulation of the RA problems is explained in this subsection.

We set x_1 to be the bit allocation vector for all the subchannels allocated to the SUs in class 1 and x_2 to be the bit allocation vector for all the subchannels allocated to the SUs in class 2. The parameters x_1 and x_2 are expressed as:

$$x_1 = [(x_{1,N}^1)^T \ (x_{1,N}^2)^T \ \cdots \ (x_{1,N}^N)^T]^T \ \in \{0, 1\}^{NK_1C \times 1} \tag{6.62}$$

$$x_2 = [(x_{2,N}^1)^T \ (x_{2,N}^2)^T \ \cdots \ (x_{2,N}^N)^T]^T \ \in \{0, 1\}^{N(K-K_1)C \times 1} \tag{6.63}$$

where $\boldsymbol{x}_{1,N}^n = [x_{1,1,n}^T \; x_{1,2,n}^T \; \cdots \; x_{1,K_1,n}^T]^T \in \{0,1\}^{K_1 C \times 1}$ indicates that subchannel n is allocated with $\boldsymbol{x}_{1,k,n} = [x_{k,n,1} \; x_{k,n,2} \; \cdots \; x_{k,n,C}]^T \in \{0,1\}^{C \times 1}$; $n = 1, \cdots, N$; $k = 1, \cdots, K_1$; C is the number of modulation schemes considered (for the model and analysis discussed in this chapter, $C = 4$). This implies that $\boldsymbol{x}_{1,k,n} = [x_{k,n,1} \; x_{k,n,2} \; x_{k,n,3} \; x_{k,n,4}]^T$. The value of $\boldsymbol{x}_2$ is arrived at in a similar manner. Then, the value $\boldsymbol{x} = \boldsymbol{x}_1 + \boldsymbol{x}_2$ is the value of the combined bit allocation vector. The mutually exclusive constraint ensures that $\boldsymbol{x}_{1,N}^n$ and $\boldsymbol{x}_{2,N}^n$ take the shape of any of the vectors $\{[0\,0 \; \ldots \; 0]^T, [1\,0 \; \ldots \; 0]^T, [0\,1 \; \ldots \; 0]^T, \ldots, [0\,0 \; \ldots \; 1]^T\}$. This implies that just one of the components in $\boldsymbol{x}_{1,N}^n$ is 1 and all other components are 0s (also true for $\boldsymbol{x}_{2,N}^n$). When $x_{k,n,c}$ is 1, it shows that the subchannel n has been allocated to SU k to transmit c bits per symbol. When $\boldsymbol{x}_{1,N}^n$ (or $\boldsymbol{x}_{2,N}^n$) has an all 0s component, it means that the subchannel n has not been allocated to any SU at all.

We define the modulation order vectors for the SUs in class 1 as $\boldsymbol{b}_1$ and for the SUs in class 2 as $\boldsymbol{b}_2$. These modulation vectors are defined as:

$$\boldsymbol{b}_1 = [(\boldsymbol{b}_{1,N}^1)^T \; (\boldsymbol{b}_{1,N}^2)^T \; \cdots \; (\boldsymbol{b}_{1,N}^N)^T]^T \in \mathbb{Z}^{N K_1 C \times 1} \tag{6.64}$$

$$\boldsymbol{b}_2 = [(\boldsymbol{b}_{2,N}^1)^T \; (\boldsymbol{b}_{2,N}^2)^T \; \cdots \; (\boldsymbol{b}_{2,N}^N)^T]^T \in \mathbb{Z}^{N(K-K_1) C \times 1} \tag{6.65}$$

where $\boldsymbol{b}_{1,N}^n = [b_{1,1,n}^T \; b_{1,2,n}^T \; \cdots \; b_{1,K_1,n}^T]^T \in \mathbb{Z}^{K_1 C \times 1}$ and $\boldsymbol{b}_{1,k,n} = [b_{k,n,1} \; b_{k,n,2} \; \cdots \; b_{k,n,C}]^T \in \mathbb{Z}^{C \times 1}$. The value of $\boldsymbol{b}_2$ is obtained in a similar fashion. Since we only considered four modulation schemes (i.e., the BPSK, 4-QAM, 16-QAM and 64-QAM), $b_{k,n} = [1\,2\,3\,4]^T$.

We define the respective data rate matrices for the two classes of SUs, $\boldsymbol{B}_i \in \mathbb{Z}^{K_1 \times N K_1 C}$ and $\boldsymbol{B}_j \in \mathbb{Z}^{(K-K_1) \times N(K-K_1) C}$ as follows:

$$\boldsymbol{B}_i = \begin{bmatrix} b_1 & b_1 & \cdots & b_1 \\ b_2 & b_2 & \cdots & b_2 \\ \vdots & \vdots & \ddots & \vdots \\ b_{K_1} & b_{K_1} & \cdots & b_{K_1} \end{bmatrix}, \quad \boldsymbol{B}_i \in \mathbb{Z}^{K_1 \times N K_1 C} \tag{6.66}$$

$$\begin{cases} b_1 = [b^T \; 0_C^T \; \cdots \; 0_C^T] \in \mathbb{Z}^{1 \times K_1 C} \\ b_2 = [0_C^T \; b^T \; \cdots \; 0_C^T] \in \mathbb{Z}^{1 \times K_1 C} \\ \vdots \quad \vdots \quad \ddots \quad \vdots \\ b_{K_1} = [0_C^T \; 0_C^T \; \cdots \; b^T] \in \mathbb{Z}^{1 \times K_1 C} \end{cases}$$

$$\boldsymbol{B}_j = \begin{bmatrix} b_{K_1+1} & b_{K_1+1} & \cdots & b_{K_1+1} \\ b_{K_1+2} & b_{K_1+2} & \cdots & b_{K_1+2} \\ \vdots & \vdots & \ddots & \vdots \\ b_K & b_K & \cdots & b_K \end{bmatrix}, \quad \boldsymbol{B}_j \in \mathbb{Z}^{(K-K_1) \times N(K-K_1) C} \tag{6.67}$$

$$
\left\{
\begin{array}{l}
b_{K_1+1} = [b^T\ 0_C^T\ \cdots\ 0_C^T] \in \mathbb{Z}^{1\times(K-K_1)C} \\
b_{K_1+2} = [0_C^T\ b^T\ \cdots\ 0_C^T] \in \mathbb{Z}^{1\times(K-K_1)C} \\
\quad\vdots \qquad\ \vdots\ \ddots \qquad \vdots \\
b_K = [0_C^T\ \ 0_C^T\ \cdots\ b^T] \in \mathbb{Z}^{1\times(K-K_1)C}
\end{array}
\right\}
$$

Since the rate weight for category one SUs is w_1 and the rate weight for category two SUs is w_2, we now write the total data rate in the objective function of Eq. (6.11) as $\max_x[(w_1 \odot b_1)^T x_1 + (w_2 \odot b_2)^T x_2]$, where $\odot$ is the Schur-Hadamard (or entrywise) product.

Now, we define $\boldsymbol{R}_I \triangleq [R_1\ \ R_2\ \ \cdots\ \ R_{K_1}]^T \in \mathbb{R}^{K_1 \times 1}$ and $\boldsymbol{R}_{II} \triangleq [R_{K_1+1}\ R_{K_1+2}\ \cdots\ R_K]^T \in \mathbb{R}^{(K-K_1)\times 1}$. Then, the constraint of Eq. (6.12), which gives the minimum data rate for the SUs in class 1, becomes $\boldsymbol{B}_i x_1 \geq \boldsymbol{R}_I$. Similarly, the constraint of Eq. (6.13), which gives the minimum data rate for the SUs in class 2, becomes $\boldsymbol{B}_j x_2 \geq \boldsymbol{R}_{II}$.

We then define a power transmission vector p such that:

$$
p = [(p_N^1)^T\ (p_N^2)^T\ \cdots\ (p_N^N)^T]^T \in \mathbb{R}^{NKC\times 1} \tag{6.68}
$$

where $p_N^n = [p_{1,n}^T\ p_{2,n}^T\ \cdots\ p_{K,n}^T]^T \in \mathbb{R}^{KC\times 1}$ and $p_{k,n} = [p_{k,n,1}\ p_{k,n,2}\ \cdots\ p_{k,n,C}]^T \in \mathbb{R}^{C\times 1}$; $p_{k,n,c}$ is the amount of power needed to transmit c bits of data for user k on subchannel n. The power constraint in Eq. (6.14) now becomes $p^T x \leq P_{\max}$.

For the underlay, overlay and hybrid CRN considerations, the interference power constraints have to be written in terms of the bit allocation vector x. In order to write the interference power constraint in Eq. (6.15) (for the underlay CRN), Eq. (6.17) (for the overlay CRN) and Eq. (6.22) (for the hybrid CRN) in terms of the vector x, we define a matrix $A \in \{0, 1\}^{N \times NKC}$ as follows:

$$
A = \begin{bmatrix}
1_{KC}^T & 0_{KC}^T & \cdots & 0_{KC}^T \\
0_{KC}^T & 1_{KC}^T & \cdots & 0_{KC}^T \\
\vdots & \vdots & \ddots & \vdots \\
0_{KC}^T & 0_{KC}^T & \cdots & 1_{KC}^T
\end{bmatrix}, \quad A \in \{0, 1\}^{N\times NKC} \tag{6.69}
$$

$$
1_{KC} = \begin{bmatrix} 1 \\ 1 \\ \vdots \\ 1 \end{bmatrix} \in \{1\}^{KC\times 1}, \qquad 0_{KC} = \begin{bmatrix} 0 \\ 0 \\ \vdots \\ 0 \end{bmatrix} \in \{0\}^{KC\times 1}.
$$

If $p \odot x$ is the Schur-Hadamard (or entry-wise) product of p and x, then $A(p \odot x)$ will be that $N \times 1$ vector in which case the nth element indicates the total power used by the nth subchannel for carrying out its data transmission. We define $\varepsilon_l \triangleq [\varepsilon_1\ \varepsilon_2\ \ldots\ \varepsilon_L]^T \in \mathbb{R}^{L\times 1}$. The constraint in Eq. (6.15), which indicates the interference power constraint for the underlay CRN, becomes:

$$H^P[A(p \odot x)] \leq \varepsilon_l, \tag{6.70}$$

while, for the overlay CRN, Eq. (6.17) is now written as:

$$H_o^P[A(p \odot x)] \leq \varepsilon_l, \tag{6.71}$$

where H_o^P is the interference gain to the PUs as a result of the SU transmission when the probabilities of miss detection and false alarm (Eq. (6.5)) are inculcated. For the hybrid network, Eq. (6.22) is written as:

$$H_o^P[A(p \odot x)] \leq \varepsilon_l, \tag{6.72}$$

By putting all the above descriptions together, the RA problem for underlay heterogeneous CRN given in Eqs. (6.11)–(6.16) can now be represented in the reformulated ILP form as follows:

$$z^* = \max_x [(w_1 \odot b_1)^T x_1 + (w_2 \odot b_2)^T x_2] \tag{6.73}$$

subject to

$$B_i x_1 \geq R_I; \ k = 1, 2, \cdots, K_1 \tag{6.74}$$

$$B_j x_2 \geq R_{II}; \ k = K_1 + 1, K_1 + 2, \cdots, K \tag{6.75}$$

$$p^T x \leq P_{\max} \tag{6.76}$$

$$H^P[A(p \odot x)] \leq \varepsilon_l \tag{6.77}$$

$$0_N \leq Ax \leq 1_N \tag{6.78}$$

$$x_1, x_2, x \in \{0, 1\} \ w_1, w_2 \in \mathbb{R}^+. \tag{6.79}$$

The RA problem for the overlay heterogeneous CRN, in the ILP form, is the same as the ILP formulation provided in Eqs. (6.73)–(6.79) EXCEPT for Eq. (6.77). In the overlay case, Eq. (6.77) becomes:

$$H_o^P[A(p \odot x)] \leq \varepsilon_l. \tag{6.80}$$

The RA problem for the hybrid heterogeneous CRN is also similar to the ILP formulation provided in Eqs. (6.73)–(6.79) EXCEPT for Eq. (6.77). In the hybrid case, Eq. (6.77) becomes:

$$H^P[A(p \odot x)] + H_o^P[A(p \odot x)] \leq \varepsilon_l. \tag{6.81}$$

6.7.3   Classical Optimisation Solutions

The reformulated RA optimisation problems for the underlay, overlay and hybrid CRN, as given in Eqs. (6.73)–(6.81), are *combinatorial ILP problems*. The reformulated ILP problems can be solved using any of the classical optimisation tools or methods that are most suited for solving ILP problems. An example of such suitable methods for solving ILP problems is the Branch-and-Bound (BnB) method. This BnB method is very adequate for solving linear combinatorial programming problems, and is employed in solving the ILP problem for RA in heterogeneous CRN developed in this chapter.

To further reduce the computational demand of the solution process, a special technique of the BnB method that is best suited to solve binary ILP problems, is employed. The special BnB technique is called *implicit enumeration* [15]. The technique of implicit enumeration is equipped with the information that each variable (in this case, the bit allocation vector x) can only take a binary number (i.e., x is either 0 or 1). It uses the binary-decision information to simplify the BnB processes of branching and bounding. It also uses the information to efficiently ascertain when a node is not going to be feasible, thereby reducing the overall computational demand of the solution process, while still not sacrificing optimality in the solution results for the CRN.

6.7.4   Other Possible Solution Methods

While we have employed the tool of studying the problem structure, alongside the tool of classical optimisation, in analysing and solving the RA problems in the heterogeneous CRN developed for the underlay, overlay and the hybrid representations, we must be clear that the solutions provided in this chapter are not necessarily exclusive or exhaustive. We emphasise that there are several other good solution tools and methods being explored by researchers for addressing their RA problems developed for the underlay, overlay and hybrid CRN. Since the RA problems for the CRN are diverse, so are the solution methods being employed to solve them.

The various solution tools and methods being used for solving RA problems in the CRN have been well discussed in the previous chapter of this book. Some of these tools, such as the heuristic and the meta-heuristic tools, give solutions that are computationally less demanding than the results from the use of the classical optimisation tools employed in this chapter. In a subsequent chapter of this book, we employ the tool of heuristics to solve a similar but expanded problem of the RA model for the CRN and compare the solutions from both the classical optimisation and the heuristic. However, we limit the solutions provided in this chapter to the ones from the classical optimisation tool. This is because it achieves optimal results for the RA problems in heterogeneous CRN. The benefits that the solution explored in this chapter have over most other solutions are its *generic nature,* *transferability, practicality, ease of analysis* and *ease of implementation* in possible real-life scenarios of the CRN.

6.8 Important Results from the Resource Allocation Modelling and Solution for Heterogeneous Cognitive Radio Networks

This section presents some important results of the generic RA model for heterogeneous CRN, as developed and analysed in this chapter. The MATLAB software is used for simulating the model. The YALMIP solver is used to carry out the optimisation. The following parameters are used for the simulations: the number of OFDMA subchannels $N = 64$, the number of PUs $L = 4$ and the number of SUs $K = 4$. Based on the heterogeneous user classifications discussed in the chapter, the SUs are categorised as: class 1 SUs $K_1 = 2$ (this represents the high-rate demand or high priority SUs, as the case may be) and class 2 SUs K_2(which is equivalent to$K - K_1$) $= 2$ (this represents the low-rate demand or best-effort service SUs). For all simulation results presented in this chapter, we used statistically independent Gaussian random variables to generate random multipath fading channels for the PUs and SUs.

Further, we set the average channel gain between the SUBS and SUs to 1 and the average channel gain between the SUBS and PUs to 0.1. We set the maximum permissible interference limit to the PUs as 0.001 mW. The interference caused by the PUs was considered as noise by the SUs. The PUs' interference to the SUs had a power spectral density of $(0.01/64)$ mW/subchannel. We used 100 randomly generated channel pairs H^s and H^p in obtaining all the simulation results. We set the BER value ρ at 0.01 for all SUs. We used a weight of unity for all the classes of SUs, except in the final results where the effects of weight were discussed.

For the results, the minimum data rate requirement for the high-rate demand class 1 SUs is 64 bits/user, while the minimum data rate requirement for the low-rate demand class 2 SUs is 32 bits/user. Generally, since the class 1 SUs have a higher data rate demand, the SUs in this class may be billed higher, or there might be some other criteria by which they are made to pay for the better QoS being provided for them. The higher priority or higher rate demand usually come at a cost for such category of SUs.

Only the results of the underlay CRN consideration are presented and discussed in this chapter. This simple reason is so as not to unnecessarily duplicate the results. In short, the results from the underlay, overlay and hybrid CRN designs follow a similar pattern and the differences are minimal. The significant differences are that, for the overlay and hybrid CRN, the interference to PUs are worse off than in the case of the underlay CRN (this is because of the effects of the interference due to probabilities of miss detection and false alarm). Therefore, in terms of the bit allocation, the results for the overlay and hybrid scenarios are slightly worse than the results obtained for the underlay CRN, the hybrid CRN performing the poorest. However, in terms of the average and total data rates, the hybrid CRN performed the best.

6.8.1   The Effects of Interference on the Bit Allocation

Figure 6.2 presents the plot of the interference channel gain between the SUBS and the PUs against the number of subchannels available in the CRN, while Fig. 6.3 presents the plot of the channel gain between the SUBS and the SUs against the number of subchannels available in the CRN for the underlay consideration. The

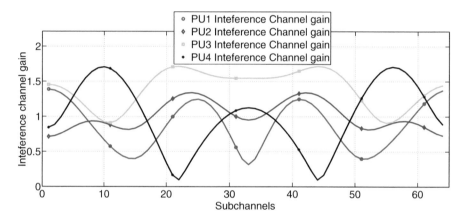

Fig. 6.2 The plot of the interference channel gain between the PUs and the SUBS against the number of subchannels

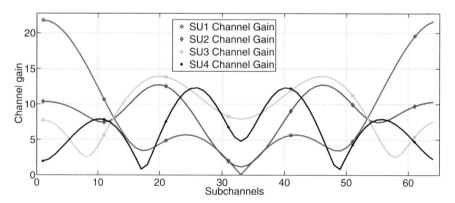

Fig. 6.3 The plot of the channel gain between the SUs and the SUBS against the number of subchannels

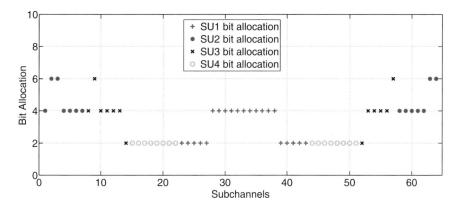

Fig. 6.4 The plot of the bit allocation for each SU against the number of subchannels

importance of the channel gain is the impact it has on the allocation of data rates to each of the SUs. The actual data rate (bits per symbol) that is allocated to each SU is plotted in Fig. 6.4. In the bit allocation plot of Fig. 6.4, an 'x' at a bit allocation of 6 for subchannel 9 simply means that subchannel 9 has been assigned to SU 3 to transmit 6 bits.

From the plots in Figs. 6.2, 6.3, and 6.4, we note the important point that the bit allocation is carried out with careful consideration of the interference gains to the PUs. When the interference gain is high (indicating that there is low or less fading), low data rates are assigned to the subchannels. The reason is that high data rates (usually from a high modulation scheme) on a subchannel will need high power, and if there is a high interference gain on such subchannel, the negative effect on the PUs will be quite significant. In the other vein, if the interference gains to the PUs are low (indicating that there is high or deep fading), assigning high data rates to the subchannels (usually from high modulation) causes minimal interference to the PUs. Therefore, the SUBS do not assign higher order modulation (e.g., the 64-QAM) to the subchannels that have high interference channel gains in order to minimise the amount of interference caused to the PUs.

It is necessary to further highlight the important idea that the RA solution model employs to help achieve optimal results in the allocation of resources for the SUs in the heterogeneous CRN. The RA model simply uses the fact that, since higher order modulation schemes usually use more power, if such modulation schemes are employed for the subchannels that have high interference channel gains to the PUs, they will cause significant harm to the PUs using those subchannels. The RA solution model therefore either completely avoids allocating data rates to those subchannels or it assigns very low data rates to those subchannels. Some examples of the use of this idea can be seen in the allocations for subchannels 2, 3, 9, 57, 63 and 64 of Fig. 6.4. In the subchannels highlighted, the data rates allocated to them are quite high. But we see that if we check in Fig. 6.2, the combined interference to the PUs on those subchannels is lower than the combined interference on the other subchannels. Conversely, the combined interference to PUs on subchannels 14–27 and 39–52 is quite high. Therefore, the data rates allocated on those subchannels are very low. This idea or principle is what the RA model employs in obtaining optimal results in the overall network utility (average data rates, throughput, etc.) for the CRN.

6.8.2 Average and Total Data Rates

Figure 6.5 is a plot of the average data rate that the SUs in each class can realise, as the interference power to the PUs is being increased. We varied the permissible interference limit of each PU, that is ε_l, between 20 and 30 dBm. The maximum SUBS power was initially set at 12 dBm but then increased to 30 dBm. From the plot, we first note that below the interference value of 20 dBm, solving the RA problem is infeasible. Next, we observe that, in the feasible region of the problem, the minimum data rate demand the class 1 SUs is achieved at all points.

Furthermore, the plot in Fig. 6.5 shows that the RA results follow a similar trend (i.e., continuous improvement) until when the permissible interference limit is about 24 dBm. After that point, the average data rate for the SUs in both classes become stable at the maximum SUBS power of 12 dBm. However, the average rate for the class 2 SUs continue to increase when the maximum SUBS power is at 30 dBm (the increase is not indefinite because if the interference limit is increased beyond the range used in this plot, the average rate will also reach its saturation point). The reason for this is that, with a higher power at the SUBS, the average data rate of the SUs improves significantly if all the other constraints remain intact. The final observation from this plot is that the RA solution model prefers to increase the average data rate of the class 2 SUs than the class 1 SUs when there is a slight

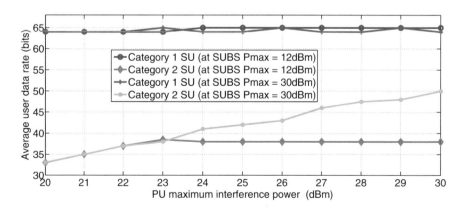

Fig. 6.5 The plot of the average data rate the SUs against the permissible interference limit to the PUs at different SUBS power

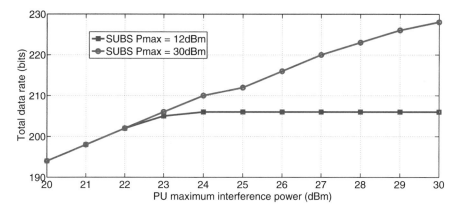

Fig. 6.6 The plot of the total data rate against the permissible interference limit to the PUs at different SUBS power

increase in resources. The reason is that it is easier to improve resource allocations to the class of SUs that have the most flexibility in their demands rather than the class of SUs that are more rigid and fixed in their demands.

Figure 6.6 presents the plot of the total data rate or throughput of the CRN against the permissible interference limit to the PUs. The permissible interference limit to the PUs is varied between 20 and 30 dBm. The maximum SUBS power at initially set at 12 dBm and later increased to 30 dBm. The result clearly shows that as the permissible interference power to the PUs is relaxed (i.e., the permissible interference power to the PUs is allowed to take higher values), the CRN achieves higher throughput. Also, we note that, at a higher SUBS power (30 dBm), the throughput keeps improving, unlike in the case of a lower SUBS power (12 dBm) where the throughput quickly stabilises, even with an increasing permissible interference limit.

6.8.3   Effects of Weight

The weight factor is one of the most important factors in the allocation of resources to different SUs or classes of SUs in the heterogeneous CRN. Indeed, the weight factor can influence the decision of the allocation model to favour some classes of SUs over other classes. Therefore, the weight factor can be used as a powerful bias mechanism in the RA decision-making process for heterogeneous CRN to provide options for further improvement that would not have been feasible or achievable if the classes of SUs do not have such weight considerations.

Figure 6.7 is a plot of the average data rate against different weight ratios for the CRN. The plot demonstrates the usefulness of weight on the amount of data rate that the different classes of SUs can achieve. We use the minimum data rate

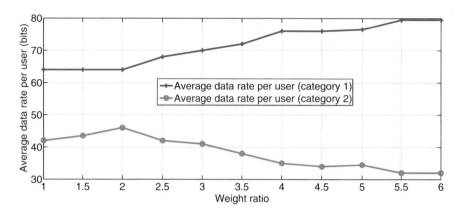

Fig. 6.7 The plot of the average data rate against different weight ratios for the different classes of SUs

classification and employ the results in Fig. 6.5 for comparison. The weight ratios between the two classes of SUs were steadily increased from unity to some higher values. From the plot, we observe that, for larger values of weight ratio, the average data rate for the SUs in class 1 increases, while the average data rate for SUs in class 2 decreases.

The implication of the results in Fig. 6.7 is that, contrary to the results obtained in Fig. 6.5, the higher weight in this case has compelled the RA model to assign higher data rates (or resources) to the SUs that has the higher demand (that is, the SUs in class 1). Without any doubt, the SUs in class 1 are the most valuable SUs. This is because, the SUs in class 1 pay a higher price, in some way, in order to get better services. It is therefore very meaningful to give them preference in the allocation of resources when there is a slight increase in the amount of resources available for the CRN. The weight factor is employed to make this happen. Despite the weight, the RA model still ensures that the minimum data rate demand for each class of SUs is still achieved in all cases, otherwise the problem will become infeasible.

Figure 6.8 is the plot that compares the performance of different weight distributions for the CRN. The work in [16] employed weights that were random numbers between 0 and 1. The weights were normalised so that the sum of all user weights became 1. To improve the results in [16], we plot three different weight distributions (the uniform, normal and exponential distributions) and compare the results. The plot shows that the normal weight distribution outperforms the exponential and uniform distributions. Of the three distributions, the uniform distribution performed the least. The plot shows that the performance of the SUs in the heterogeneous CRN could be slightly influenced by the choice of the weight distributions that is used in the network design.

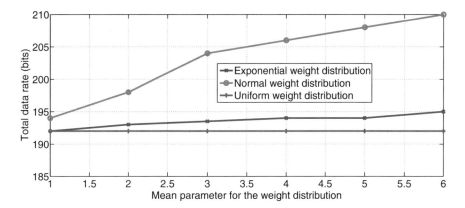

Fig. 6.8 The plot of the total data rate against the mean parameter for different weight distributions

6.9 Summary of the Chapter

This chapter has developed, analysed and discussed significant RA solution models for the heterogeneous CRN. The RA models for the heterogeneous CRN discussed in this chapter are generic models that fit the modern and practicable CRN designs. The RA models are developed for the underlay, overlay and hybrid CRN. To achieve optimal or very-close-to-optimal solutions for the RA problems, the optimisation tool of studying the problem structure is employed in realising an ILP reformulation of the NP-hard problems. The reformulated ILP problems are then solved by the use of the tool of classical optimisation. Some useful results are presented and discussed to show the relevance of interference, user classification and weights on the overall performance of the heterogeneous CRN.

References

1. B.S. Awoyemi, B.T. Maharaj, A.S. Alfa, Resource allocation for heterogeneous cognitive radio networks, in *Proceedings of IEEE Wireless Communications and Networking Conference* (2015), pp. 1759–1763
2. Y. Rahulamathavan, K. Cumanan, L. Musavian, S. Lambotharan, Optimal subcarrier and bit allocation techniques for cognitive radio networks using integer linear programming, in *Proceedings of 15th IEEE Workshop on Statistical Signal Processing* (2009), pp. 293–296
3. B. Awoyemi, B. Maharaj, A. Alfa, Optimal resource allocation solutions for heterogeneous cognitive radio networks. Digit. Commun. Netw. **3**(2), 129–139 (2017). http://www.sciencedirect.com/science/article/pii/S2352864816301043
4. R. Xie, F. Yu, H. Ji, Dynamic resource allocation for heterogeneous services in cognitive radio networks with imperfect channel sensing. IEEE Trans. Veh. Technol. **61**(2), 770–780 (2012)
5. S. Wang, M. Ge, C. Wang, Efficient resource allocation for cognitive radio networks with cooperative relays. IEEE J. Sel. Areas Commun. **31**(11), 2432–244 (2013)

6. B.S. Awoyemi, B.T. Maharaj, A.S. Alfa, QoS provisioning in heterogeneous cognitive radio networks through dynamic resource allocation, in *Proceedings of the IEEE AFRICON* (2015), pp. 1–6

7. C. Shi, Y. Wang, P. Zhang, Joint spectrum sensing and resource allocation for multi-band cognitive radio systems with heterogeneous services, in *Proceedings of IEEE GLOBECOM* (2012), pp. 1180–1185

8. B.S. Awoyemi, B.T. Maharaj, A.S. Alfa, Resource allocation in heterogeneous buffered cognitive radio networks. Wireless Commun. Mobile Comput. **2017**, 7385627, 1–12 (2017)

9. Y. Rahulamathavan, S. Lambotharan, C. Toker, A. Gershman, Suboptimal recursive optimisation framework for adaptive resource allocation in spectrum-sharing networks. IET Signal Process. **6**(1), 27–33 (2012)

10. M. Pischella, D. Le Ruyet, Cooperative allocation for underlay cognitive radio systems, in *Proceedings of the 14th IEEE Workshop on Signal Processing Advances in Wireless Communications* (2013), pp. 245–249

11. S. Wang, Z.-H. Zhou, M. Ge, C. Wang, Resource allocation for heterogeneous cognitive radio networks with imperfect spectrum sensing. IEEE J. Sel. Areas Commun. **31**(3), 464–475 (2013)

12. F. Chen, W. Xu, Y. Guo, J. Lin, M. Chen, Resource allocation in OFDM-based heterogeneous cognitive radio networks with imperfect spectrum sensing and guaranteed QoS, in *Proceedings of the 8th International Conference on Communications and Networking in China* (2013), pp. 46–51

13. H. Peng, T. Fujii, Hybrid overlay/underlay resource allocation for cognitive radio networks in user mobility environment, in *2013 IEEE 78th Vehicular Technology Conference (VTC Fall)* (2013), pp. 1–6

14. T. Xue, X. Dong, Y. Shi, Resource-allocation strategy for multiuser cognitive radio systems: location-aware spectrum access. IEEE Trans. Veh. Technol. **66**, 884–889 (2017)

15. W.L. Winston, M. Venkataramanan, *Introduction to Mathematical Programming*, 4th edn. (Thompson Brooks/Cole, Pacific Grove, 2003)

16. P. Cheng, Z. Zhang, H. Huang, P. Qiu, A distributed algorithm for optimal resource allocation in cognitive OFDMA systems, in *Proceedings of the IEEE International Conference on Communications* (2008), pp. 4718–4723

Part III
New Directions in the Development of Cognitive Radio Networks

The field of cognitive radio networks is still very much a new and evolving field of modern technology. As cognitive radio networks evolve, some recently-introduced and highly-significant analytical tools (such as queueing theory) and techniques (such as cooperative diversity) are being explored in order to achieve an improved overall network performance and productivity for cognitive radio networks. Furthermore, new works are focussing on how cognitive radio networks are interacting with and/or influencing some of the other newly-developing technologies such as the fifth-generation and the internet-of-things networks. In some cases, the influence of cognitive radio networks on such other emerging technologies are already well established while in other cases, the impact is not yet fully exposed. In this part of the book, new concepts that explore possible/ongoing interactions between cognitive radio networks and other emerging technologies are discussed. Some useful and exciting ideas for further development and eventual implementation of the cognitive radio network are proposed.

Chapter 7
Queuing Systems in Resource Allocation Optimisation for Cognitive Radio Networks

7.1 Queuing-Related Problems in Heterogeneous Cognitive Radio Network

In typical cognitive radio networks (CRN) scenarios, the primary users (PUs) of the network usually have the upper hand in the access and usage of the resources, especially the spectrum resource. This is because the PUs are likely to be the licensed owners of the spectrum in which the SUs are designed to co-operate. As a result, the secondary users (SUs) in the secondary network operate under the stringent conditions or constraints imposed on them by the primary network. To achieve optimal results, therefore, the secondary networks in the CRN must develop and devise mechanisms by which they can overcome the effects of the strict conditions imposed on them by the primary network, while still achieving their communication goals [1, 2].

The crippling conditions and/or constraints on secondary networks in the CRN become exacerbated when the CRN is developed as heterogeneous systems. In most heterogeneous CRN designs, the demands of one SU may differ from the demands of another SU, or one category of SUs may have different demands to the demands of other categories of SUs. In good network designs, therefore, the different demands of the SUs or SU categories must be efficiently and timeously met [3, 4]. For this to be feasible, there is a high demand to develop resource allocation (RA) models for the CRN that will incorporate the peculiar properties and dynamics of the different heterogeneous categories of SUs into their CRN designs [5]. Similarly, better solution models for the RA problems in the CRN that can achieve optimal allocation of the limited CRN resources in a manner that is fair and favourable to all categories of SUs are a new necessity [6, 7].

As modern RA models and solutions for heterogeneous CRN are being developed and analysed, an important condition that must be considered is that different SUs or SU categories have different levels of delay tolerance for them to meet an acceptable quality-of-service (QoS) requirement. In most cases, the different SUs

B. TJ Maharaj, B. S. Awoyemi, *Developments in Cognitive Radio Networks*, https://doi.org/10.1007/978-3-030-64653-0_7

or SU categories usually have differing delay tolerance characteristics. The time delay characteristics of SUs or SU categories in the CRN are often dependent on the kind of services being provided by the SUs. Therefore, the delay tolerance levels for different SUs can actually be a useful criterion for separating the SUs in the heterogeneous CRN into different classes or categories for proper analyses and optimal resource usage.

Since the various SUs in the heterogeneous CRN do have different delay tolerance levels, the CRN can be designed to leverage the time delay characteristics in providing improved services for all SUs. To achieve this, the SUs can be categorised and serviced based on their delay tolerance characteristics, for instance. This would then imply that, if resources are not immediately available, the SUs or SU categories that do not require or demand immediate services could use a buffer (or queue) to store their data. Those SUs or SU categories must then wait, in most cases, for some acceptable duration of time, until there are enough resources being provided for them to carry out or complete their data transmission.

To achieve optimal resource usage and management in heterogeneous CRN designs, it is important to know the instances and durations of time delays that are acceptable for different SUs or SU categories in the CRN. This knowledge, alongside the knowledge of the characteristics of the queues resulting from such delays, are essential in achieving the QoS requirements of the SUs, and in providing optimal or near-optimal solutions for the RA problems in heterogeneous CRN. The aspect and concept of queues in telecommunication networks are well addressed by studying and applying queuing theory or queuing systems. Queuing theory is a powerful tool that can be employed for studying and analysing RA for heterogeneous CRN, especially when time delays and buffer considerations are a part of the CRN design.

Apart from time-delay problems, there are other problems in the RA optimisation for heterogeneous CRN that are best captured and addressed using queuing descriptions and analyses. Such problems include *user prioritisation*, *resource sharing*, *data buffering*, etc. A good study on queuing is therefore very necessary to study queuing theory or queuing systems for RA in the CRN. This chapter discusses useful ideas, models and practises of queuing theory for solving RA problems in heterogeneous CRN. Some useful examples and analyses of the use of queuing theory models to solve RA-related problems in heterogeneous CRN are also well discussed in this chapter.

7.2 Description of Queuing Theory for Resource Allocation in Cognitive Radio Network

A *queuing system* involves one or more *customers* (which could take the form of persons, systems, connections, users, packets, etc.) that need some form of *service* (which could be in form of transaction, transmission, access, etc.) being served by one or more *servers* designed to render such services. A *queuing model* encapsulates the probability distribution of the inter-arrival times of customers, the probability distribution of times to render services to customers, the number of servers that are available to render services at any particular time and the queue capacities (if the queue is assumed to have a finite length) of the queuing system [8]. Queuing theory (systems and models) is a well-established tool for studying, analysing and implementing new, advanced and/or improved technological designs, especially in science and engineering.

For the CRN, queuing theory and the analyses of queue characteristics can help improve the RA performances of new CRN designs. To achieve this, appropriate queuing models must be developed for the CRN. The appropriate queuing model for heterogeneous CRN must incorporate buffers for the different SUs or SU categories since different SUs (or category of SUs) have different delay priorities or delay profiles. If such models are developed and properly analysed using queuing theory, the delay characteristics of SUs in the CRN may be exploited in realising significant improvements in the RA solutions for the heterogeneous CRN.

7.3   Queuing-Based Resource Allocation Solutions for Cognitive Radio Network

Most recent heterogeneous CRN systems are designed in such a way that multiple SUs can be allowed to transmit in a single subchannel at different time. The RA solutions for heterogeneous CRN must determine the optimal approach for assigning individual SUs to particular subchannels for their data transmission at definite instances of time, and for specific time durations, so as to be able to realise the best productivity for the CRN. These are indeed RA optimisation or optimisation-related problems and they are best treated as such.

As already discussed in a previous chapter, there are a number of optimisation tools and methods that have been and are being developed and employed for carrying out RA in the CRN. These RA solutions have been quite useful in describing and signifying how the subchannels available in the CRN should be allocated to the heterogeneous SUs for them to transmit their data, for how long such transmissions should take place, etc. While the tool of optimisation has been most useful in solving the RA problems in the CRN, however, it seems that the performances of a good number of RA solutions and solution models can still be further improved. The tool of queuing theory is one useful tool that is being incorporated into the optimisation solution models in order to help achieve further improvements in the RA performances of the CRN. In this chapter, some very useful applications of queuing theory in achieving improved RA solutions for heterogeneous CRN are discussed.

7.4   Queuing Model for Multi-Modal Switching Service Levels

For optimal resource utilisation in the CRN, the possibility of a hybrid design that combines both the overlay and underlay architectural designs have been proposed (see Chap. 2). The hybrid CRN is a rather complex CRN design. In the hybrid CRN, if the PUs are unavailable to use their subchannels, the SUs operate in the overlay mode. In the overlay mode, the SUs are allowed to transmit their data using entire resources such as high transmission rates, maximum transmission power, etc. However, immediately the PUs are back to use their subchannels, the SUs must switch to the underlay mode. In the underlay mode, the SUs transmit at a lower resource level (lower transmission power, modulation, etc.) so that they do not cause significant interference to the PUs. During underlay transmission, the SUs must still be able to achieve signal-to-noise ratio (SNR) values that meet some QoS requirements. The work in [9] is a classic example of the use of queuing theory in achieving the needed inter-switch between the different possible modes in a hybrid CRN. The work further studied how the switching activity between different levels or modes of spectrum access can affect the overall performance of the RA solutions for the CRN.

7.4.1   Different Modes as Different Service Levels

In the work in [9], being a hybrid CRN system, the SUs are made to operate in different or distinct modes. These different modes are indicative of the different 'service levels' that the SUs can operate by. In the overlay mode, the SUs operate at the highest transmission rate. In the underlay mode, the SUs operate at different service levels of lower transmission rates. The SUs can switch from one service level to another service level, depending on the prevailing circumstances of the CRN.

The discrete-time single-server queuing models are useful models for analysing networks in which there are different service levels being provided by the same server. Individual probability distributions are used to describe the service time of each service mode. This interesting idea of servicing the SUs in different modes is equated to analysing queues with *working vacations* [10]. In the case considered in [9], when the server is at full service, the overlay mode, which is the mode with the highest service rate, is activated. All the other working rates of the server, which correspond to the different service levels in the underlay mode, are the services that are activated during the periods of the working vacation.

7.4.2   Description of the System Model for Multi-Modal Switching Service Levels

The system model in [9] represents the case of the CRN with multiple SUs attempting to transmit to a base station using a common channel that is licensed to a PU. A single server queue with infinite capacity is used to describe the model. In the model, the arriving units are the SUs. The single server represents the channel access for the SUs. The PU has the highest service priority. After the PU, the SU at the head of the queue has the next priority of service. The first-come-first-serve (FIFO) scheme is employed in servicing the SUs. The scheduling of channel access for the SUs in the queue is carried out by a base station.

In the network, a discrete-time process is used to describe the queuing model employed. In discrete-time processes, equally spaced discrete time points are used to represent the various events and state transitions in the system. Non-negative integer numbers $0, 1, 2, \ldots$ are used to indicate the discrete time points. The SUs arrive in discrete time with their workloads (i.e., their data for transmission) according to a *general distribution*. Furthermore, in this single server system, the *discrete Markovian arrival process* (MAP) is used to describe the inter-arrival times of the SUs' data. Sub-stochastic matrices D_0 and D_1 of dimensions $n \times n$ are used to establish the discrete MAP events. The authors explained that this discrete MAP is best for modelling the correlation between the inter-arrival times. In that case, the elements in D_0 represent the transitions between the transient states when there is no arrival and the elements in D_1 represent an arrival event which leads to an instantaneous restart of the process into one of the transient states.

In the model developed in [9], the server could operate at different transmission rates with more than one service modes. Therefore, the total time used to provide services for the SUs is dependent on the spectrum that is activated for use in each mode, as the server interchanges the modes. To help capture these behavioural changes in the model, a distinction is made between the time already used up and service time left for each unit that arrives, subject to the condition that service rates may change when the service modes change in the course of servicing the SU.

A discrete phase-type (PH) distribution is used to model the workload (W) of the SU that has just arrived into the system. Similarly, the work habit (H) of the server is modelled using the discrete PH distribution. The value of H is a reflection of the stochastic processing time required to carry out a unit of work. Both W and H are discrete random variables with finite supports. This implies that the service time (S) that will be required to satisfy an SU that has just arrived, having a workload W, will have a PH distribution of the order that is obtained by multiplying the orders of W and H.

Further, in the analysis of the model presented in [9], one important aspect that has been neglected in most works on hybrid architectural designs of the CRN, that is, the aspect of the changing power levels of the PU as they transmit their data, is incorporated. This means that the PU's transmission power levels P_p can change in the course of the period of data transmission by the SUs. Therefore, the power level at which the SU must transmit P_{st} would also have to be dynamic. The implication of this is that, in a particular session, the rate of data transmission for the SU may change before the session elapses. To capture this reality, a multi-modal service level for the SUs is developed. This further extends the construction of the PH distribution of S.

7.4.3 Analysis and Performance Results of the Multi-Modal Switching Model

In the analysis of the model developed in [9], the total number of service modes in the system is represented by N. Therefore, $n = 0, 1, 2, \ldots, N - 1$ represents all the different modes that are available or applicable to the system. Any SU that arrives into the system starts being serviced in one of the N modes, as long as the server is not being used at that time and no other SUs are waiting for service. If these conditions are not met, the SU that has just arrived will have to wait and move in the queue until it gets to the top of the queue. To make the model complete and accurate, there is a further $(N + 1)$th mode that is applicable. Because of the method employed in analysing the model, since the first mode is the mode $n = 0$, then this last $(N + 1)$th mode will be the mode $n = N$. In this mode $n = N$, the server is incapable of attending to any SU, making it impossible for the SUs to transmit at any power level in the underlay mode. The channel is simply being fully used by the PU, and therefore, there is a zero service rate to the SUs.

Putting all of the possible modes of transmission together, it means that when the SU transmits its data in mode $n = 0$, it transmits at the highest mode with the highest power level possible (i.e., it is in the overlay mode with the PUs being unavailable). If mode $n = 0$ is not applicable, it may use any of the other modes $n = 1; 2; \ldots; N - 1$. This translates to different lower modes with lower power levels for the SUs to transmit their data, while the PU is also using that channel (i.e., the underlay mode). In general, when the SU uses a mode/power level n, it achieves a better data transmission than when it uses the mode/power level $n + 1$. Therefore, $H_n \leq H_{n+1}; n = 0, 1, 2, \ldots, N - 1$. If the SU is in mode $n = N$, this is the worst case where the channel conditions do not allow the SUs to transmit any data alongside the PUs. Once an SU joins the queue, it requires a total processing time S to complete its service after it starts to be served.

The model is analysed using the discrete-time Markov chain (DTMC) queuing model. By assuming that the system is stable, it was possible to carry out the steady-state analysis of the system. From the steady-state analysis, results on the distribution of the number of SUs in the system at different times were obtained. Some other important performance measures, such as the mean queue length and the mean number of SUs in the system, were also obtained. The results showed that, on average, the SUs that initiate their services in a low mode (such as $n = 4$ or 5) usually require more time to complete the data transmission than those that initiate their services in a high mode (such as $n = 0$ or 1). This is because of the lower service rates that are obtainable at low service modes. The eventual consequence of carrying out data transmissions using the lower modes is that the SUs' data are backlogged for a longer time, which may cause longer queue lengths and higher service times. Therefore, more SUs prefer to initiate their services in higher modes because of the better service rates that they can achieve. The results from the model can be employed in devising a good policy for making decisions for the SUs to help optimise their performance and to significantly reduce their 'costs' of operation as they switch from one service mode to other service modes in modern CRN applications.

7.5   Queuing Model for Increased Spectrum Utilisation

The work in [11] used the tool of queuing theory or system to devise a mechanism for achieving *channel sharing*, whereby two SUs can occupy and share the same channel simultaneously with the PU, in order to further improve resource utilisation in the CRN. When channel sharing is employed, the interference to the PU as a result of the combined transmission of the SUs must still be within the permissible temperature (power) limit of the PU. This means that the SUs must share the available transmission power, which must still be below the PUs' threshold interference. Some network rules are employed to allocate power to each SU to use for their respective data transmission. In the rules, more power is allocated to the SU that has the higher priority. If the second SU is absent, the single SU can use a transmission power that is as high as the permissible interference threshold of the PU.

7.5.1   System Model and Analysis of the Queuing Model for Increased Spectrum Utilisation

The system is modelled using the underlay CRN architecture. While the PU operates, the model used the weighted *head of line processor sharing technique* to prioritise how the two SUs are to share the resources for an optimal network experience. The system was analysed using the M/M/1 queuing model. Two possible implementations of the model are considered, namely, the pre-emptive case and the non-pre-emptive case.

In a pre-emptive case, the class that arrives must pre-empt the service of the class that is already being served. In the context of the secondary network for the underlay CRN that was considered in [11], the pre-emptive case means that if there is only one SU that is using the channel and the other SU arrives, the transmission power of the initial SU is immediately adjusted so as to accommodate the SU that has just arrived. The implication is that the power that will be used to complete the service of the SU that was already in the system is going to be reduced. There are two classes of SUs considered in the CRN model developed, namely the high priority (HP) SUs and the low priority (LP) SUs. An infinite number of HP SUs is allowed in the CRN system whereas a finite number of LP SUs is permitted in the system. A buffer of magnitude K is used to limit the number of LP SUs. This means that, at any particular time, the CRN only allows a maximum of $K + 1$ LP SU packets in the system. The well-used first-come-first-serve queue discipline is employed for the network.

The mean arrival rate of the entire CRN system is obtained by adding the mean arrival rates for the two SU categories. There are four different mean service rates that were defined. The first one is the mean service rate of the HP queue when there is no LP SU present. The second one is the mean service rate of the LP queue when there is no HP SU present. The third one is the mean service rate of the HP queue when there are LP SUs present. The final one is the mean service rate of the LP queue when there are HP SUs present. The model is analysed using the state space approach for obtaining the steady-state responses for the system. The transition probability matrix is seen to have a quasi birth death (QBD) process structure. Furthermore, the R-matrix obtained was shown to have a somewhat special structure, which was that the R-matrix had repeating rows and columns. This special structure of the R-matrix was said to be due to the fact that queue arrivals do not depend on the state of the other queue and as a result, the packets in the LP queue could reduce even though there were still HP packets in the system.

In a non-pre-emptive case, the class that arrives does not need to pre-empt the service of the class that is already being served. In the context of the secondary network for the underlay CRN that was studied in [11], the non-pre-emptive case means that if there is only one SU that is using the channel and the other SU arrives, the transmission power of the initial SU will not be immediately adjusted to accommodate the new SU that has just arrived. This implies that, until transmission is completed, there will be no interruption to the transmission power being used in the service of the SU packet that was already in the system. It is only after the completion of the service of the SU packet that is currently being served that the transmission power can be adjusted to now serve both categories of SUs, if they both have packets in their queues. The shared rate will continue to be employed to serve the two SUs until one SU completes its transmission. The queue of the SU that is remaining will then be served at the maximum transmission power until a packet of the other SU arrives again. This is applicable to the HP SUs and LP SUs. Similar to the pre-emptive case, an infinite number of HP SUs is allowed in the system while a buffer of size K is used for the finite LP SUs, implying a maximum of $K + 1$ LP SU packets in the system at any given time.

The state space analysis for the non-pre-emptive case was more complex than in the pre-emptive case. This is because, at every point, the analysis had to capture the type of SU that was currently being served while a new SU arrived. To achieve this, the authors used dummy states to keep track of the packet that is already in service on arrival of a packet to the other queue. Steady-state analysis of the system was carried out using the state space approach. Similar to the pre-emptive case, the transition probably matrix also had a QBD process structure. However, unlike in the pre-emptive case, because of the dummy states that were included in the non-pre-emptive case, the elements of the resulting R-matrix were made up of smaller matrices themselves, making the problem more difficult to solve. Despite the complexity, some simulation results were obtained for the network.

7.5.2   Performance Results of the Queuing Model for Increased Spectrum Utilisation

The authors in [11] did carry out some simulation tests to evaluate the performance of the CRN system in consideration. The results show that in all cases, the queuing model introduced through the channel share priority scheme benefited both queues (i.e., the queues of both SUs or category of SUs that were present in the CRN system being considered) in the important aspect of spectrum access and utilisation. More so, the queue of the LP SUs benefited the most from the channel sharing scheme that was incorporated into the CRN design. In most occasions, there were huge improvements in the normal transmission rates of the SUs to the maximum allowable transmission rates for the network. The improvements were more pronounced when the traffic intensity on the queue of the HP SUs was low. The work therefore helps to establish the importance of queuing theory as a great tool for improving resource utilisation in different heterogeneous CRN considerations.

7.6   Queuing Model for Heterogeneous Users with Different Delay Profiles

Another important work that had demonstrated use of queuing theory or system for improving RA solutions in the CRN is the work in [12]. The work incorporated the possibility of the SUs having different delay considerations in the development and analysis of an RA model for heterogeneous CRN. The authors used the demands of the SUs to place them into different classes. Each class had its own queue or buffer and its own service capacity. Each SU was placed into a queue by considering its distance from the secondary user base station (SUBS). The assumption was that the SUs were mobile, and therefore, they could move from one queue to another. Additionally, for optimal results in the RA process, it was possible to move some demands from one queue to another queue if there was a chance of getting served better in the new queue. This resulted in significant improvements in the network performance for the CRN.

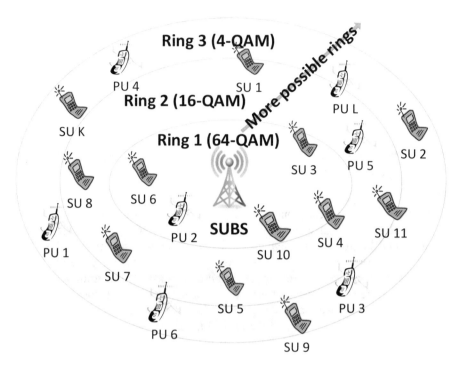

Fig. 7.1 A description of the system model for heterogeneous buffered CRN

7.6.1 System Model of the Heterogeneous Buffered Cognitive Radio Network

The system model of the heterogeneous CRN developed in [12] is shown in Fig. 7.1. The heterogeneous CRN that was considered in the work was a centralised underlay CRN design. This meant that the SUs were permitted to transmit using all of the spectrum space of the PUs, as long as they did not violate the permissible interference limit of the PUs. In the model, two classes of SUs were considered, namely, the delay-sensitive (DS) SUs and the delay-tolerant (DT) SUs. The amount of delay time that was acceptable to achieve the desired quality of service (QoS) was used to differentiate the DS SUs from the DT SUs. All the SUs had mobility characteristics. They were also able to change their modulation and coding schemes in a dynamic manner because they were equipped with the adaptive modulation and coding (AMC) technique.

Furthermore, each SU was placed in a virtual ring. The ring to which an SU belongs was dependent on its distance from the SUBS. The ring that was closest to the SUBS employed the highest AMC technique for its operation. The ring that was farthest from the SUBS employed the lowest AMC technique for its operation. There was a queue or buffer for each ring. All the data transmission requests of the

SUs in a particular ring were placed in a queue of that ring. The network subchannels were then used to carry out the data transmission or service provisioning for the SU requests. The queues acted as a buffer that kept the transmission requests of the SUs which were not immediately attended to for an acceptable delay time until resources were available to attend to such requests.

7.6.2   Model Analysis of the Heterogeneous Buffered Cognitive Radio Network

In the CRN model developed in [12], there are N subchannels, corresponding to the number of parallel servers in each ring (or queue). Each queue was finite and had a maximum length of Y. Since the SUs were mobile, it was possible to adjust the arrival rates into queues in order to achieve the best productivity for the CRN. To adjust the arrival rates into queues, some part of the DT requests were moved from a farther ring (queue) to a closer ring (queue). This made it possible to transmit such data demands at a higher rate. This arrangement helped to reduce the energy and time needed to transmit the data, which resulted in meaningful improvements in the capacities of the SUs in the CRN.

For the CRN model developed in [12], the queuing analysis was carried out to particularly emphasise the importance of the fraction of demands that was moved from one queue to another queue. For clarity and ease of representation, just two concentric rings were analysed. In that case, there were just two parallel queues but more than one server (subchannels) could serve each queue. However, more than two rings could also be used and the analysis would still be applicable. The ring that was nearest to the SUBS employed the 64-QAM modulation technique (this meant that it used 6 bits per symbol) and the ring that was farthest to the SUBS employed the 4-QAM modulation technique (this meant that it used 2 bits per symbol). The Poisson distribution was used to model the arrivals into the queues. The arrival rate for queue 1 was λ_1 while the arrival rate for queue 2 was λ_2. An exponential distribution was used to model the service for the queues. The service rate for queue 1 was μ_1 while the service rate for queue 2 was μ_2. The values of μ_1 and μ_2 were equivalent to the data rates of the AMC techniques being operated in rings 1 and 2, respectively. This meant that 6 bits per symbol was achievable for queue 1 and 2 bits per symbol was achievable for queue 2.

Since the rate of service for queue 1 was higher than for queue 2, the service per unit time was much faster in queue 1 than in queue 2. Also, since both DT and DS demands arrived into each queue, and users were mobile, the productivity of the CRN may be improved if some of the DT demands of a farther ring were moved to a closer ring. The model therefore established the value of the fraction of the DT demands of queue 2, that is θ, that should be moved to queue 1 so that a higher transmission rate could be achieved. The greatest problem solved in [12] was to obtain an optimal value for θ. This value corresponded to the best fraction of the demands to be moved from one queue to another queue in order to realise the optimal resource usage for the CRN.

The queuing model used to analyse the system was the continuous-time Markov chain (CTMC) queue with a finite buffer. From the model, the sum of initial arrival to queue 1 and the fraction of arrival to queue 2 that was brought to queue 1 gave the total arrival into queue 1. The total arrival into queue 1 was thus $\lambda_1 + \theta\lambda_2$. In essence, the total arrival to queue 2 was the remaining part of the initial arrival to queue 2 minus the part that had been moved to queue 1, that is, $\theta\lambda_2$. If there was no portion of queue 2 that was transferred to queue 1 or if there were no arrivals at all into queue 2, then $\theta\lambda_2 = 0$ and arrival to queue 1 just remained λ_1.

The model was analysed by using the state space approach to obtain steady-state responses for the system. Equilibrium balance equations for the developed CRN system were obtained by applying the steady state conditions. The balanced system was solved using the standard M/G/1 queue analysis. The 'M' indicated that the arrival was the standard memoryless distribution while the 'G' meant that the service followed a hyper-exponential distribution (i.e., the combination of two exponential distributions). The arrival of packets was combined into a single arrival stream. Furthermore, to obtain optimal values for the parameter θ and to study its effects, the queuing system was solved using the state reduction approach. With the state reduction approach, the value of θ was easily varied while adjusting the arrival rates into each queue. This made it easy to evaluate the effects of θ on the overall network performance of the CRN.

The state reduction approach also made it possible to obtain the equilibrium probabilities of the Markov chain that developed in the course of the analysis of the model. With the equilibrium probabilities, it became possible to obtain the optimum value of the parameter θ by applying the Newton's method of numerical analysis. The authors noted that the parameter θ is not limited in definition and application to only mean the fraction of the DT demands that is transferred from one queue to the other. There may be other definitions or applications for the parameter θ. It could actually be defined to be any other factor that may be used to determine the relation between one set of users and another. For instance, if user priorities are used in classifying and/or categorising the SUs, θ could be a fraction of the higher priority demands. The important point is that, by defining and obtaining the value of θ that optimises resource utilisation, the overall productivity of the heterogeneous CRN can be greatly improved.

7.6.3 Performance Results for the Heterogeneous Buffered Cognitive Radio Network

Some important performance measures were investigated in [12] to show the effect of θ on the CRN. Two particular performance measures were investigated, namely, blocking probability and system throughput. Both performance measures were obtained from the steady-state probabilities. Several other performance measures, such as the average number of packets in the queue or in the system, could be easily obtained from the blocking probability and system throughput. The results showed that the blocking probability decreased for an increasing value of θ, while the throughput increased for an increasing θ value. This is important because it indicated that the overall performance of the CRN was improved through the queuing model incorporated.

Further results showed, however, that the performance of the CRN did not increase indefinitely with an increasing θ value. Rather, at some point, the improvement in performance was completely eliminated and a gradual reduction in performance began to take place. This happened because the continuous transfer of the DT demands in queue 2 to queue 1 in anticipation of a better service resulted in a tipping point, which corresponded to the optimum θ value. After this point, any further increase in the amount of queue 2 demands that was transferred to queue 1 did not result in an improvement in the performance of the CRN. In fact, beyond the optimum value of θ, there was a significant increase in the blocking probability. This was because the data in queue 1 became too large, eventually causing the overall productivity of the CRN to decrease. It is therefore clear from the analysis and results of the model in [12] that only by moving the right amount of data requests from one queue to another using appropriate queuing models can the desired improvement in the allocation and utilisation of the limited resources in the heterogeneous CRN be realised.

7.7 Performance Evaluation of Queuing-Related Resource Allocation Solutions in Cognitive Radio Networks

We have already established that the newly developing CRN systems are being designed to serve heterogeneous multiple users, especially in the secondary networks. Moreover, the SUs in the secondary network must be capable of sharing the diverse spectrum bands with the PUs in the network. The manner in which the multiple PU channels are assigned to the SUs can affect or impart the QoS that is realised or achieved for the CRN. Optimisation, especially when applied with queuing theory, has been shown to be one of the most powerful tools for achieving the best results in the RA for heterogeneous CRN.

As more and more solution models that incorporate queuing theory into RA optimisation for the CRN are being developed, it is becoming very difficult but highly necessary to establish means by which the various RA solutions can be compared, so as to evaluate the performances of the various models and approaches being developed for achieving optimal RA for the CRN. However, despite its importance, the works on performance evaluation of different queuing-based solution models for RA in the CRN have been quite few. We present some recent/ongoing efforts that discuss joint evaluating and comparison of performance measures for RA solutions in the heterogeneous CRN.

In designing performance methods and models for measuring and comparing RA solutions in the CRN, it is necessary to develop and employ methods or models that are *configurable*. For a performance method to be configurable, it means that the analytic model and the results of the performance evaluation carried out are not necessarily fixed while employing or applying that performance method. In other words, the method is such that a number of the input parameters can be easily adjusted so that the performance measures can be fitted to meet different requirements.

Unfortunately, most RA solutions for the CRN have not employed very configurable performance evaluation methods in their designs. The implication of non-flexibility in the RA approaches and solutions is that, if there are significant or even slight changes in the input parameters (such as a change in the channel conditions), the performance of the CRN may become unpredictable because the allocation methods employed in the design are simply non-configurable and as such, they cannot adapt to such changes in network conditions or requirements. It is therefore necessary to investigate and develop configurable performance methods to properly study, evaluate and compare RA solutions for the heterogeneous CRN.

7.8   Performance Framework for Queuing-Based Resource Allocation in Cognitive Radio Network

The works in [13–15] are recent attempts at developing configurable queuing-based frameworks for modelling, studying and evaluating RA performance of heterogeneous CRN. In such frameworks, various environmental parameters and CRN settings can be jointly considered and/or are incorporated into the CRN system. We use the work in [14] as a base framework for our study and analysis. In the spectrum sensing section of the framework, imperfect spectrum sensing is assumed, which easily incorporates the perfect spectrum sensing conditions. The framework developed a spectrum sensing model to help understand and interpret the relationship between the sensing outcome of the SUs and the occupancy states of the PU channels. The parameters that indicate the activities of the PUs and the parameters that indicate the results of the spectrum sensing activities by the SUs are built into this sensing model.

Furthermore, in the section of the framework that deals with channel allocation, the allocation procedure is modelled as a Markov process that combines the spectrum sensing model and a newly proposed flexible channel allocation protocol (called the distribution probability matrix (DPM) protocol) in carrying out the channel allocation. In the section of the framework that deals with data transmission, an adapted AMC technique is employed. A truncated automatic repeat request (ARQ) technique is incorporated into the AMC technique to improve the adaptation in the modulation and coding scheme. The work then developed an analytic procedure to describe how the framework can adapt to different conditions, while still achieving impressive results in its data transmission using the AMC technique incorporated with the truncated ARQ scheme. The works used the ideas of queuing modelling to establish how the framework can evolve and be analysed, and for deriving various performance measures for the CRN.

7.8.1   System Model and Analysis of the Performance Framework

The performance framework in [14] is developed using the centralised or infrastructure-based CRN architecture. The SU network operates within the coverage range of the PU network, but each network is controlled by its own base station. Information exchange (channel conditions, spectrum sensing results, synchronisation, etc.) between the base stations of the SUs and the PUs is assumed to be reliable. In each time slot, all the physical layer frames in the PU network and the SU network are synchronised and they have similar time duration. The frequency band authorised to the PU network is divided into subchannels. These PU subchannels are also used by the SUs in an overlay architectural design. Each SU's mobile device stores its data packets in a finite buffer.

Furthermore, in describing the PU activities, the highly used first-order Markov process with two states is employed. This means that for each PU, transmission slot is either 'busy' or 'free'. The time slot of an SU is divided into three successive parts, namely, the *spectrum sensing* part, the *channel allocation* part and the *data transmission* part. In the sensing part, the possibility of imperfect spectrum sensing is assumed. Therefore, the probabilities of false alarm and miss detection are incorporated. In the part that addresses channel allocation for the CRN, the SU base station assigns all the subchannels that have been sensed to be 'free' to the SUs, while using the flexible and configurable DPM protocol that has been developed for this purpose. In the part that addresses data transmission for the CRN, the model employed the AMC technique, alongside the truncated ARQ technique, to achieve the data transmission for the network.

The truncated ARQ technique works on the principle of positive feedback. If a data packet is successfully transmitted, the transmitter receives an acknowledgement message from the receiver. If the data packet is not successfully transmitted, the receiver sends back a non-acknowledgement message. This prompts the transmitter to retransmit the data packet. A data packet that has been retransmitted for a number of consecutive times and is yet unsuccessful is dropped by the transmitter. Once the data transmission process has been completed, all the data packets that arrived during that time slot are placed in the buffer. The network rejects all the data packets that arrive after the buffer is full.

The tool of queuing theory/system is employed to analyse the RA performance framework proposed for heterogeneous CRN. The queuing analytical framework developed simultaneously evaluates individual users in the system. The framework establishes the link between various channel allocation protocols and performance measures for the CRN. The framework is very useful for analysing the impact of system settings and environmental parameters on the optimal allocation results and network performance of the heterogeneous CRN.

7.8.2   Benefit of a Performance Framework

The important advantage that the performance frameworks have over other means of evaluating RA performances for the CRN is that all the activities and processes considered in a framework are configurable to adapt to various possible situations and/or scenarios of the CRN. Hence, even if the CRN that is being considered or investigated is not exactly the same as the one that was employed in the framework, the framework may still provide near-accurate estimate results for the RA problem of the CRN being considered. This is possible because the framework will help fit the different settings of the new CRN design to the parameters in the framework model and will generate appropriate results for the CRN in consideration.

7.9   Performance Implications of Queuing-Based Resource Allocation in Cognitive Radio Networks

In the relatively new analytical framework developed in [14] and others for studying the performance measures of the RA solutions in the CRN, the important goal is to design the network such that the configurable components, such as the *arrival process*, the *service process* and the *allocation protocol*, are flexible to adapt to as many situations or scenarios as may be required. With this framework, it is easy to obtain numerical results for the typical parameters that are mostly used in the literature for studying RA problems in the CRN. The results from the framework can also be used to determine how the performance measures of a particular SU will change if the environmental parameters and settings are altered in a CRN system.

7.9.1   Important Performance Measures

Several performance measures to evaluate and compare RA solutions for the CRN can be easily obtained using the framework developed in [14] and [15]. The most important performance measures that can be used to study RA solutions for the CRN are the *gross throughput* of the CRN, *average queue length*, the *packet rejecting rate*, the *average packet delay*, the *packet collision rate* and the *packet dropping rate* for SUs in the network. Furthermore, with the framework, some useful generalisations on the CRN can be derived. For instance, one of the most interesting results obtained by using the framework, which also confirms an intuitive concept about the CRN, is that if a particular SU has a higher chance of being allocated to PU subchannels, the average number of packets in the buffer will decrease because the SU has a better chance of transmitting its packets. Other such important generic observations on the performance of the CRN can also be drawn by employing frameworks that are versatile, configurable and reliable.

7.9.2   Implications of Proper Performance Evaluation

From the results obtained using the framework proposed in [14] and [15], a number of important information on the network performance of RA solutions for the CRN are deduced. The most important implications of the performance measures and evaluations achieved using the framework developed in [15] for obtaining RA solutions for the CRN are described as follows:

- In the design of the CRN, if there are certain limitations on some of the performance measures in consideration (say for instance, that the data rate cannot be lower than a certain threshold value), a good framework can help to obtain the boundary of distribution probability of the SU that the channel allocation protocol must apply in order to guarantee an effective CRN realisation.
- A good framework can be used to optimise one or more performance measures for the CRN. For instance, in the framework developed in [15], the relationship between DPM parameters and other performance measures (such as the average number of collision packets, the average number of packets in the buffer and the average number of rejected packets) can be used to carry out the optimisation.
- In the situation where the CRN is made to share a number of subchannels or even the subchannels from a different kind of system with different parameters, the allocation protocol that is employed in the framework can be easily altered to meet the performance requirements or to simply optimise the overall performance of the CRN.
- The relationship that is established between the performance measures and the settings of the CRN can be used to analyse and identify the most important CRN settings that can influence the performance measures the most. Moreover, by employing the framework, the CRN can be better designed to meet some specific selected performance requirements of the CRN.
- In designing channel allocation schemes for the CRN, there is usually a trade-off between the performance to be considered and the complexity of the network. A good framework can help in determining and studying the important complexity-performance trade-off for practical applications, and to assist in providing the needed information for QoS decision making in the CRN.

7.10   Summary of the Chapter

This chapter has presented and discussed the concept of queuing theory and queuing systems as a powerful tool for achieving improved RA solutions, particularly for modern heterogeneous CRN. Furthermore, the tool of queuing theory and systems (which incorporates queuing models and analyses) can be well employed for addressing the problems of time delays, buffering, fairness in resource sharing, etc. in the RA optimisation for heterogeneous CRN. Finally, the introduction of new performance frameworks that combine both the tools of optimisation and queuing theory in designing configurable performance measures and procedures for the RA solutions in heterogeneous CRN is a positive development in the design and implementation of modern CRN.

References

1. B.S. Awoyemi, B.T. Maharaj, A.S. Alfa, Resource allocation in heterogeneous cooperative cognitive radio networks. Int. J. Commun. Syst. **30**(11), e3247 (2017). dac.3247. https://onlinelibrary.wiley.com/doi/abs/10.1002/dac.3247
2. L. Wang, W. Xu, Z. He, J. Lin, Algorithms for optimal resource allocation in heterogeneous cognitive radio networks, in *Proceedings of the 2nd International Conference on PEITS*, vol. 2 (2009), pp. 396–400
3. B.S. Awoyemi, B.T. Maharaj, A.S. Alfa, QoS provisioning in heterogeneous cognitive radio networks through dynamic resource allocation, in *Proceedings of the IEEE AFRICON* (2015), pp. 1–6
4. M. Kaplan, F. Buzluca, A dynamic spectrum decision scheme for heterogeneous cognitive radio networks, in *Proceedings of the 24th International Symposium on ISCIS* (2009), pp. 697–702
5. B.S. Awoyemi, B.T.J. Maharaj, A.S. Alfa, Solving resource allocation problems in cognitive radio networks: a survey. EURASIP J. Wirel. Commun. Netw. **2016**(1), 176 (2016). https://doi.org/10.1186/s13638-016-0673-6
6. L. Zheng, C.W. Tan, Cognitive radio network duality and algorithms for utility maximization. IEEE J. Sel. Areas Commun. **31**(3), 500–513 (2013)
7. B. Awoyemi, B. Maharaj, A. Alfa, Optimal resource allocation solutions for heterogeneous cognitive radio networks. Digital Commun. Netw. **3**(2), 129–139 (2017). http://www.sciencedirect.com/science/article/pii/S2352864816301043
8. F. Palunčić, A.S. Alfa, B.T. Maharaj, H.M. Tsimba, Queueing models for cognitive radio networks: a survey. IEEE Access **6**, 50801–50823 (2018)
9. A.S. Alfa, H.A. Ghazaleh, B.T. Maharaj, A discrete time queueing model of cognitive radio networks with multi-modal overlay/underlay switching service levels, in *2018 14th International Wireless Communications Mobile Computing Conference (IWCMC)* (2018), pp. 1030–1035
10. A.S. Alfa, *Queueing Theory for Telecommunications*. LLC (Springer Science+Business Media, New York, 2010)
11. H.M. Tsimba, B.T. Maharaj, A.S. Alfa, Increased spectrum utilisation in a cognitive radio network: an m/m/1-ps queue approach, in *2017 IEEE Wireless Communications and Networking Conference (WCNC)* (2017), pp. 1–6
12. B.S. Awoyemi, B.T. Maharaj, A.S. Alfa, Resource allocation in heterogeneous buffered cognitive radio networks. Wirel. Commun. Mob. Comput. **2017**(7385627), 1–12 (2017)

13. S. Wang, B.T. Maharaj, A.S. Alfa, Resource allocation and performance measures in multi-user multi-channel cognitive radio networks, in *2016 IEEE 83rd Vehicular Technology Conference (VTC Spring)* (2016), pp. 1–5

14. S. Wang, B.T. Maharaj, A.S. Alfa, Queueing analysis of performance measures under a new configurable channel allocation in cognitive radio. IEEE Trans. Vehi. Technol. **67**(10), 9571–9582 (2018)

15. S. Wang, S. Maharaj, A.S. Alfa, A virtual control layer resource allocation framework for heterogeneous cognitive radio network. IEEE Access **7**, 111605–111616 (2019)

Chapter 8
Cooperative Diversity for Resource Optimisation in Cognitive Radio Networks

8.1 The Problem of Interference in Cognitive Radio Networks

As already well established, cognitive radio networks (CRN) promises to help mitigate or resolve the problem of spectrum scarcity for new and emerging next-generation (xG) wireless communication networks [1, 2]. The improvement in spectrum efficiency and system throughput, as promised through the primary-secondary networking arrangement in the CRN, provides an enabling platform for achieving xG wireless communication capabilities. With this huge promise, the CRN is being developed as one of the key technologies to drive near-future telecommunication possibilities [3, 4]. While there are still some challenges with the CRN, a lot of research works on the CRN continue to go on and new breakthroughs are emerging for some of the challenges of the CRN that have already been identified.

One of the main challenges of the CRN that still require a lot of work to help resolve is the problem of network interference [5]. Interference concerns in the CRN are mostly of the situation that the secondary users (SUs) should not be allowed to cause undue amount of interference to the primary users (PUs) when the SUs carry out their communication. Even still, the possibility of the SUs causing undue interference among themselves is surely a worthy cause for concern. The reason is that network interference—any and all forms of it—can negatively affect the performance of the CRN in many undesirable ways.

In most CRN set ups, the primary network has priority over the secondary network in the use of resources and other things. Therefore, if the primary network has stringent interference conditions, the secondary network productivity may become very low. For instance, in the underlay architecture, even though the SUs can use all the frequency band of the PUs, their transmission power much be low enough such that the amount of interference to the PUs is very bearable [6]. In cases where the PUs are highly sensitive or they have incredibly low permissible interference thresholds, it becomes a big challenge for the CRN to realise the utmost for the network. And just as the underlay architecture has interference issues, the overlay architecture has significant interference issues too, as the problems of miss detection and false alarm in the overlay CRN designs are both interference-related problems.

In the likely CRN scenarios or situations where the amount of interference that the PUs permit during the data transmission of the SUs is very low, the total data rate or network capacity that the CRN achieves can be negatively affected [7]. This problem of permissible interference can, in fact, cast doubts on the CRN's worthiness, and/or on whether or not the ongoing investments in the CRN would be a worthwhile pursuit, unless the interference problem is adequately addressed. Solving the interference problem in the CRN is thus an important aspect of its design and implementation.

8.2   Attempts at Solving the Interference Problem in Cognitive Radio Networks

Indeed, there are recent/ongoing efforts to address and solve the interference problem in the CRN. In [7], for instance, to help ameliorate the effects of high interference received by the PUs as a result of the SUs' data transmission, the authors proposed to use a power control estimation mechanism that is adaptive, distributed and neighbour coordinated. With this mechanism, the SUs can sense when the PUs are sending their signals and then avoid the use of the subchannels being used by the PUs in carrying out their transmission. This ensures that the possibility of harmful interference reaching the PUs is minimal. The major challenge with this arrangement is that it does not exactly solve the interference problem, rather, it seeks to avoid it. Hence, the productivity of the CRN may still be very low, because the SUs only carry out their data transmission at the periods when the PUs are not using their spectrum, just as in most other regular overlay arrangements.

In the work in [8], some models that jointly consider pre-coding in the physical layer and channel allocation in the medium access layer to help limit the amount of interference that the SUs can cause to the PUs when they transmit their signals are developed and investigated. The authors then proposed some distributed cross-layer algorithms to maximise the throughput of the network, while also minimising the interference between the PUs and the SUs in the system. The problem with the suggested solution, however, is that the already complex RA formulations for the CRN became even more complex by the cross-layer optimisation solution being proposed.

The work in [9] developed a model that simultaneously optimises both the channel assignment and the transmission power control for RA in the CRN. The authors did claim that because of the joint optimisation, the effects of both the adjacent-channel interference between the PUs and the SUs, as well as the co-channel interference among the SUs, are mitigated. However, this claim cannot be substantiated as there was no interference mitigation technique that was employed for the network.

Importantly, in the works mentioned above and in many other similar works that have attempted to address the challenge of interference in the RA problems and solutions for the CRN, the aspect of heterogeneity has been mostly ignored. We have shown in previous chapters of this book that the concept of heterogeneity is very important in providing RA formulations for the CRN that are realistic and practicable. This was also established in [10]. Therefore, the limited solutions that have been proposed still leaves the problem of interference in the CRN as open-ended, and one that requires new and urgent solutions. The work in [5] is a good example of ongoing works in this regard.

8.3   Cooperative Diversity Approach to Solving the Interference Problem in Cognitive Radio Networks

In previous chapters of this book, we did show that, in order to achieve optimal or very-close-to-optimal solutions to the RA problems in the CRN, the best practice would be to allocate low data rates (or even zero data rate) to the subchannels that have high PU interference profiles and high data rates to the subchannels that have low interference profiles. This is a reasonable practice. The reason is that, if high data rates are allocated to the subchannels that have significant interference profiles, the SUs will use high modulation schemes and transmission power on those subchannels. This will result in a high amount of interference being caused to the PUs due to the high interference gains on those subchannels [11, 12]. This important decision of not allocating high data rates to the subchannels with high PU interference profiles has been shown to have greatly increased the throughput and productivity of the CRN.

However, even with those important decisions and rules for minimising possible interference in place, the productivity that can be achieved for the CRN is still limited. The reason is that, with such rules, there would be a number of subchannels which, due to the possible high interference channel gains to the PUs, are either not allocated at all or are allocated to only carry out data transmission at data rates that are quite low. Therefore, if the goal is to improve the productivity of the CRN further, it will require that new and/or better channelling procedures that can help actualised the possibility of high data rates for almost all the subchannels that are available in the secondary network of the CRN, while still not causing significant interference to the PU network, are investigated.

This chapter introduces and investigates *cooperative diversity* as a new and highly promising solution for the interference limitation in the CRN. Cooperative diversity has been recently proposed to achieve an improved wireless channel conditioning by employing diversity gains among spatially dispersed users [13, 14].

The improvement in wireless channel conditioning using cooperative diversity happens when the cooperating users (also referred to as relays or nodes) unit to form a virtual multiple-input multiple-output (MIMO) arrangement. In other words, just as it happens in conventional MIMO systems, the cooperating users in a cooperative diversity setting, though in different locations, use their antennas in helping each other transmit (or retransmit) their data to a given destination. This usually brings about sizeable increases in the capability and reliability of the communication system.

Even though it is still a relatively new communication tool, there are already a fairly impressive number of cooperative diversity methods that have been and/or are being developed and applied for xG wireless communication systems. The most commonly used cooperative diversity methods are the *amplify-and-forward* method, the *decode-and-forward* method, the *store-and-forward* method and the *coded cooperation* method [15].

In terms of relay-selection classification, there are a number of cooperative diversity categories that have been described. If the cooperative diversity is classified based on the number of cooperators that are selected, we may have *single-relay cooperation* or *multiple-relay cooperation*. If the cooperative diversity is classified based on whether the cooperation actually happens or not, we may have *opportunistic cooperation* or *incremental cooperation* [14]. Regardless of the method and/or category of cooperative diversity employed, what is important is that at the destination, there is an impressive improvement in the network capacity as a result of the better signal quality that is realised when cooperative diversity is employed.

In the remaining parts of this chapter, we developed and studied the use of cooperative diversity for mitigating the challenge of interference in the RA for the CRN. We show that, by addressing the interference challenge using cooperative diversity, the resourcefulness and overall productivity of the CRN can be made to be much better. This is very significant for practical CRN realisation.

8.4   Recent Works on Cooperative Diversity for End-2-End Communication in Cognitive Radio Networks

In recent times, there have been attempts to introduce cooperation diversity into the RA problems and solutions in the CRN. Some good examples of recent works that have developed models that describe useful cooperation between SUs, so as to realise better resource utilisation for the CRN are found in [3, 16–19]. In [3] and [16], some relays that employ the decode-and-forward method are selected to help the SUs in the CRN. In order to make the resulting optimisation problem solvable, the allocation of the subchannels that are available to the SUs is first carried out. The subchannel allocation to the USs was based on the channel gains of each SU, and the possible interference that it will cause to PUs. After the subchannels have been allocated to the SUs, an appropriate transmission power is then assigned to each subchannel.

In the model developed in [17], similar to the models investigated in [3] and [16], the decode-and-forward cooperative method is employed to help the SUs in transmitting their data. The cooperative diversity incorporated helped in the improvement of the throughput of the CRN. The RA optimisation problem that was developed was shown to be non-convex. However, the RA problem was solved by employing the methods of dualisation and decomposition. The RA problem was decomposed into two parts, namely relay assignment and power allocation, and solved using the dualisation approach.

In the work in [18], the authors used a primary decomposition method to solve the RA problem that was developed for cooperative CRN. The RA problem was formulated as a power allocation problem and solved by splitting the convoluted problem into individual power allocation problems. In the work in [19], the authors jointly maximised the sum rate of all the SUs and the PUs, as the SUs cooperatively carry out their data transmission. The model first carried out subchannel allocation to the SUs before assigning transmission power to the SUs and the PUs in an iterative manner. This helped to achieve results that are very close to the optimal values that can be achieved.

The works mentioned above are some of the most recent works that have incorporated some kind of cooperative diversity in their design of the CRN to help achieve some form of objective, which all gears toward improving the productivity realised by the network. An important observation is that *the cooperative diversity introduced in most works on the CRN are not designed to address the problem of interference to PUs*. In other words, the interference challenge is still there, despite the attempts to employ cooperative diversity to help improve network performance in the CRN. Thus, the aspect of addressing interference using cooperative diversity in the CRN is still a significant research gap that is open for a lot more.

In this chapter, the cooperative diversity design that is introduced seeks to address and mitigate the limiting effects of the problem of PU interference in the CRN. The cooperation model first takes care of the interference problem before resources are allocated to the SUs. Therefore, through cooperation, the negative effects of the interference to the PUs are ameliorated, making it possible to realise significant improvements in the resource utilisation for the heterogeneous cooperative CRN.

8.5 A System Model for Cooperative-Based Resource Optimisation in Cognitive Radio Network

This section develops and describes a generic system model to study the application of cooperative diversity to help mitigate the effects of interference in the CRN. A generic heterogeneous CRN model has been developed and analysed in a previous chapter of this book. That model is extended in this chapter to incorporate cooperative diversity in the RA problem formulation and solution for the CRN. The resulting CRN design is referred to as heterogeneous cooperative CRN.

The system model of the new heterogeneous cooperative CRN is made up of K heterogeneous SUs and L similar PUs. All the SUs and the PUs fall within the coverage area of the secondary user base station (SUBS). The total number of heterogeneous OFDMA subchannels available for the SUs is N. The K heterogeneous SUs have different demands and priorities. The SUs are classified as K_1 SUs and $(K - K_1)$. The K_1 SUs have a minimum rate requirement while the $(K - K_1)$ SUs are the best effort service users. The demands of the SUs in category one are attended to first since they have a minimum rate requirement. The category one SUs are therefore the high priority SUs. The SUs in category two are simply the best effort users. The SUs in category two therefore only share whatever resources are left after the category one SUs have been attended to, using a fair proportional rate constraint. A slow fading model is used to model the environmental conditions of all the subchannels.

Importantly, as the CRN communicates, the heterogeneous cooperative CRN that is developed must decide if it is to use *direct communication* or *cooperative communication*, depending on the immediate network condition. What influences the decision on whether direct or cooperative communication is employed is the *interference to the PUs*. Direct communication will be employed if the potential interference to the PUs is minimal on a particular link. However, if by employing direct communication on a link, a high amount of interference to PUs is observed which would potentially limit the entire CRN productivity, then, the direct link is ignored and the cooperative communication is employed on that link. A description of the system model for heterogeneous cooperative CRN is shown in Fig. 8.1.

The particular cooperative diversity method that is being employed to mitigate the effects of interference in the RA solution for the CRN is the *incremental, single-relay selection* cooperative diversity method. The incremental part of the cooperative diversity method implies that cooperation is only employed when it is necessary. The single relay-selection part of the cooperative diversity method implies that a single 'best' relay is selected to help achieve the cooperation. This particular method of cooperative diversity is employed because it is more feasible and it results in a minimal network overhead. Any SU that demands to cooperate in order to achieve data transmission to the SU at the destination terminal (D) is called the source SU (SSU).

It is necessary to reiterate the condition for which cooperative communication is employed. The important condition that triggers cooperative communication is that direct communication between the SSU and D is not feasible because of high interference channel gain to the PUs. As a result of the possible high interference, the allocating algorithm would not have allocated any subchannel to the SSU, or, at best, it would have allocated a very small number of subchannels to the SSU to transmit at a very low data rate. This is the particular problem that the cooperative diversity method being incorporated into the CRN attempts to address.

During cooperation, therefore, the SSU identifies and contacts a cooperating SU (CSU) that has good channel conditions (both from the SSU to the CSU, as well as from the CSU to the D SU) and poor interference channel gain to the PUs. The new cooperating channel is then employed to carry out data transmission, thereby circumventing the possibility of interference to the PUs. Thus, the limiting effects of poor channel conditions are well mitigated using cooperative diversity. In the next section, we describe the method for identifying and choosing the best relay (that is, the CSU) from among the other SUs in the CRN.

8.6   The Relay-Selection Process in Cooperative Diversity for Cognitive Radio Networks

The system model of the heterogeneous cooperative CRN described in Fig. 8.1 uses the SUBS as its communicating hub while operating a centralised control system arrangement. We assume that the SUBS communicates with the SUs perfectly. Each SU estimates and communicates its channel condition and its PU interference gain to the SUBS. There are no restrictions on which SU can be appointed as the CSU for another SU, but rather, each SU has an equal chance of being appointed. The decision on whether or not a SU requires a cooperator, and on which cooperator

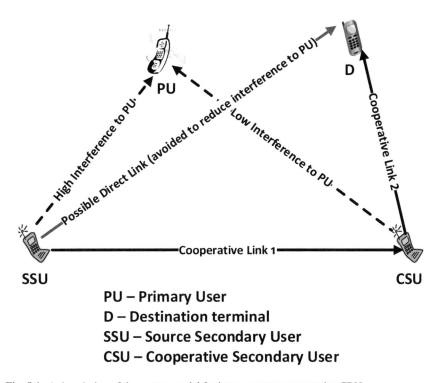

PU – Primary User
D – Destination terminal
SSU – Source Secondary User
CSU – Cooperative Secondary User

Fig. 8.1 A description of the system model for heterogeneous cooperative CRN

is most appropriate or best for a particular SU, is made by the SUBS. The SUBS
also makes contact with the best SU that is selected, and assigns it as the CSU. The
assumption is that, at the time of cooperation, the SU that is designated as the CSU
is free and not transmitting its own data. The information on whether a particular
SU is transmitting its own data or not, alongside the possible level of interference to
the PU and the estimated channel condition are jointly transmitted to the SUBS via
the control channel by each SU.

The SUBS makes the decision on which SU is to be selected as the CSU and
informs the SU that it has been selected to be the CSU. The SUBS makes its decision
by considering the SU with the *best channel condition* and the *least interference
gain to the PU*. Once the selected SU accepts its nomination, its choice as the CSU
is immediately sent to the SSU by the SUBS. All the other SUs are not contacted,
and as such, they simply continue with their normal transmission (or, if they were
not at all busy, they just maintain their idle state for that period). The SSU sends its
data to the CSU, which is then forwarded to the D SU on the subchannels that have
been assigned for them. Each time frame for data transmission is divided into two
time slots. The SSU sends its data to the CSU in the first time slot while the CSU
sends its data to the D SU in the second time slot.

To determine the joint channel condition of the SSU and the CSU, we let $H_{k,n}^s$ be
the channel gain between the SU k and the SSU, which is the SU that is selected to
be the CSU, on subchannel n. We let $H_{k,n}^r$ be the channel gain between this CSU and
the D SU on the nth subchannel. In the first slot, the SSU sends its data to the kth
SU (the CSU) on subchannel n with transmission power $P_{k,n}^s$. In the second slot, the
kth relay (CSU) sends its data to the D SU on subchannel n using the transmission
power $P_{k,n}^r$. Then, we calculate the data rate on each transmission slot as [3]:

$$
\begin{aligned}
c_{k,n}^s &= \log\left(1 + \frac{P_{k,n}^s |H_{k,n}^s|^2}{\sigma_r^2 + \sum_{l=1}^{L} J_{k,n}^l}\right), \\
c_{k,n}^r &= \log\left(1 + \frac{P_{k,n}^r |H_{k,n}^r|^2}{\sigma^2 + \sum_{l=1}^{L} J_n^l}\right)
\end{aligned}
\tag{8.1}
$$

where σ_r^2 and σ^2 are the variance of the noise at the kth relay (CSU) and the D SU,
respectively.

In the same vein, we denote $J_{k,n}^l$ as the interference to the kth relay (CSU) by the
lth PU on subchannel n. We denote J_n^l as the interference to the D SU by PU l on
subchannel n. The receivers of the CSU and D SU measure these interference values
which takes them as noise. We note that the effective data rate when cooperative
communication is employed, $c_{k,n,C}$ is usually not higher than the value of the
minimum of the data rates in the two hops. Therefore,

$$
c_{k,n,C} = \min\left(c_{k,n}^s, c_{k,n}^r\right).
\tag{8.2}
$$

In the instances when there is no need for cooperative communication, the
SSU communicates directly with the D SU using the subchannels that have been
allocated to them. In that case, the data rate is $c_{k,n,D}$. The actual value of the

data rate c on a subchannel that uses either cooperative or direct communication depends on the type of modulation that is used on that subchannel. We consider that the network employs four types of modulation schemes, namely the binary phase shift keying (BPSK), 4-quadrature amplitude modulation (QAM), 16-QAM and 64-QAM. The respective data rates for the four modulation schemes being considered are $c = 1, 2, 4$ and 6 bits per OFDMA symbol. Given that a particular value of the bit error rate (BER) ρ is to be realised, the BPSK modulation requires a minimum transmission power of $P(c, \rho) = N_\phi [c \times erfc^{-1}(2\rho)]^2$ (where $c = 1$). Similarly, for the M-ary QAM, the minimum transmission power that is required is given as $P(c, \rho) = \frac{2(2^c - 1)N_\phi}{3}[erfc^{-1}(\frac{c\rho\sqrt{2^c}}{2(\sqrt{2^c-1})})]^2$ ($c = 2, 4$ or 6 for the 4-QAM, 16-QAM and 64-QAM, respectively) where $erfc(x) = (\frac{1}{\sqrt{2\pi}})\int_x^\infty e^{\frac{-t^2}{2}} dt$ is the complementary error function, $\pi = (22/7)$, and N_ϕ is the noise power spectral density. The value of N_ϕ is taken to be the same value for all the subchannels.

To maintain a particular ρ value, an increase in the number of bits on a subchannel will result in a non-linear increase in the amount of transmission power required to communicate on that subchannel. We obtain the minimum transmission power $P_{k,n}(c_{k,n}, \rho)$ needed to transmit $c_{k,n}$ bits on subchannel n for SU k by dividing the power $P(c_{k,n}, \rho)$ of the SU k on subchannel n by the channel gain $H_{k,n}^c$ between the SUBS and the SU k on subchannel n. This power is given as:

$$P_{k,n}(c_{k,n}, \rho) = \frac{P(c_{k,n}, \rho)}{H_{k,n}^c}. \tag{8.3}$$

8.7 Problem Formulation of Resource Allocation in Heterogeneous Cooperative Cognitive Radio Network

In analysing the RA for the heterogeneous cooperative CRN being studied, we denote R_k as the minimum data rate of the SU k in category one of the secondary network, and γ_k as the normalised proportional fairness factor for each SU in category two of the secondary network. The data rate R_i indicates the achievable data rate for the element i. We denote $\Phi_n = \Sigma_{k=1}^K P_{k,n}$ as the maximum power on the nth subchannel, where $P_{k,n}$ is the transmission power of the SU k on the subchannel n ($P_{k,n,C}$ represents the power employed for cooperation communication, while $P_{k,n,D}$ represents the power employed for direct communication). We represent the interference power gain matrix between the PU and the SUBS as $H^P \in R^{L \times N}$. Then, the vector $H_{l,n}^P$ stands for the subchannel interference power gain between the PU l and the SUBS on subchannel n ($H_{l,n,C}^P$ represents the gain matrix for cooperation communication, while $H_{l,n,D}^P$ represents the gain matrix for direct communication).

We denote ε_l as the maximum amount of interference that the PU l can permit from all the SUs that are transmitting their data. We denote P_{max} as the maximum transmission power of the SUBS. We denote $X_{k,n,D}$ as a binary (0, 1) variable used to restrict each subchannel to either employ direct communication or cooperative communication. This binary variable ensures that each subchannel uses either direct or cooperative communication, but not both. The formulation of the RA problem for the heterogeneous cooperative CRN is now given as:

$z = \max$

$$\sum_{n=1}^{N} \left(\sum_{k=1}^{K_1} \left[X_{k,n,D} c_{k,n,D} + (1 - X_{k,n,D}) c_{k,n,C} \right] \right.$$

$$\left. + \sum_{k=K_1+1}^{K} \left[X_{k,n,D} c_{k,n,D} + (1 - X_{k,n,D}) c_{k,n,C} \right] \right);$$

$$\times c_{k,n,D}, c_{k,n,C} \in \{0, 1, 2, 4, 6\} \tag{8.4}$$

subject to

$$\sum_{n=1}^{N} (c_{k,n,D} + c_{k,n,C}) \geq R_k; \quad k = 1, 2, \cdots, K_1 \tag{8.5}$$

$$\frac{R_k}{\displaystyle\sum_{i=K_1+1}^{K} R_i} = \gamma_k; \quad k = K_1 + 1, K_1 + 2, \cdots, K \tag{8.6}$$

$$\sum_{n=1}^{N} \left(\sum_{k=1}^{K} \left[X_{k,n,D} P_{k,n,D} + (1 - X_{k,n,D}) P_{k,n,C} \right] \right) \leq P_{max} \tag{8.7}$$

$$\sum_{n=1}^{N} \Phi_n H_{l,n,D}^{P} \leq \varepsilon_l; \quad l = 1, 2, \cdots, L \tag{8.8}$$

$$\sum_{n=1}^{N} \Phi_n H_{l,n,C}^{P} \leq \varepsilon_l; \quad l = 1, 2, \cdots, L \tag{8.9}$$

$$c_{k,n,D} = 0 \; if \; c_{k',n,D} \neq 0, \; c_{k,n,C} = 0 \; if \; c_{k',n,C} \neq 0,$$

$$\forall k' \neq k; \; k = 1, 2, \cdots, K \tag{8.10}$$

$$X_{k,n,D} \in \{0, 1\}, \; X_{k,n,D} = 1 \; if \; c_{k,n,D} \neq 0$$

$$X_{k,n,D} = 0 \; \text{otherwise.} \tag{8.11}$$

Equation (8.4) captures the objective function for the RA problem. In this case, the objective function is indicative of the sum throughput or total data rate that is realised by all the SUs in the network using both direct and cooperative communication. The constraint in Eq. (8.5) is the minimum data rate constraint. This constraint ensures that the minimum data rate demand for each SU in category one is met. The Equation in (8.6) is the constraint which ensures that the service of the category two SUs is at best effort. We used a proportional fairness factor to share the remaining resources among the category two SUs. The constraint in Eq. (8.7) ensures that the total transmission power of all the SUs, both at direct and cooperative transmission, do not exceed the maximum transmission power of the SUBS. The constraint in Eq. (8.8) is to ensure that the amount of interference that reaches each PU when the SUs are using direct communication do not exceed the set interference threshold value. Just like the constraint in Eq. (8.8), the constraint in Eq. (8.9) ensures that the set interference limit is not exceeded during cooperative communication. The constraint in Eq. (8.10) is the mutually exclusive constraint. The constraint ensures that only one SU is assigned to each subchannel. Therefore, once we have allocated subchannel n to the SU $k' \neq k$, the data rate for subchannel n must be 0 for any other user k.

The constraint in Eq. (8.6) can be changed to:

$$R_k = \gamma_k \times \sum_{i=K_1+1}^{K} R_i,$$

where $\sum_{i=K_1+1}^{K} R_i$ is the addition of all the data rates for all the category two SUs. If we represent $\gamma_k \times \sum_{i=K_1+1}^{K} R_i$ by $\tilde{\gamma}_k$, Eq. (8.6) is better written as:

$$R_{K_1+1} : R_{K_1+2} : \ldots : R_K = \tilde{\gamma}_{K_1+1} : \tilde{\gamma}_{K_1+2} : \ldots : \tilde{\gamma}_K. \tag{8.12}$$

8.8   Optimal Solution for the Resource Allocation Problem in Heterogeneous Cooperative Cognitive Radio Networks

Very clearly, we see that the RA problem formulation for the heterogeneous cooperative CRN given in Eqs. (8.4–8.11) is not a linear programming problem. A simple way to ascertain this is to note the non-linearity of the power constraint in Eq. (8.7). However, such RA problems can still be solved, as already well discussed in a previous chapter of this book. One of the tools of optimisation that was described in that previous chapter, that is, the tool of studying the structure of the RA problem, is employed in this chapter to optimally solve the RA problem for heterogeneous cooperative CRN. Particularly, the method of reformulation is employed to change the RA problem into an integer linear programming (ILP) problem. The reformulated problem is easy to solve using an appropriate classical optimisation technique. The reformulation process is quite similar to the one carried out in a previous chapter of this book.

We set x_I to be the bit allocation vector for all the subchannels that are allocated to all the category one SUs (both for the direct and the cooperative communication, such that $x_I = (x_{I,D} + x_{I,C})$). We set x_{II} to be the bit allocation vector for all subchannels that are allocated to all the category two SUs (both for the direct and cooperative communication, such that $x_{II} = (x_{II,D} + x_{II,C})$). The parameters x_I and x_{II} are expressed as:

$$x_I = [(x_{I,N}^1)^T \ (x_{I,N}^2)^T \ \cdots \ (x_{I,N}^N)^T]^T \ \in \{0, 1\}^{NK_1C \times 1} \tag{8.13}$$

$$x_{II} = [(x_{II,N}^1)^T \ (x_{II,N}^2)^T \ \cdots \ (x_{II,N}^N)^T]^T \ \in \{0, 1\}^{N(K-K_1)C \times 1} \tag{8.14}$$

where $x_{I,N}^n = [x_{I,1,n}^T \ x_{I,2,n}^T \ \cdots \ x_{I,K,n}^T]^T \ \in \ \{0, 1\}^{KC \times 1}$ indicates that the subchannel n has been allocated to a SU in category one with $x_{I,k,n} = [x_{k,n,1} \ x_{k,n,2} \ \cdots \ x_{k,n,M} \ ]^T \ \in \{0, 1\}^{C \times 1}; \ n = 1, \cdots, N; \ k = 1, \cdots, K; \ M$ is an indication of the number of modulation schemes that are being used (in this case, $M = 4$). This implies that $x_{I,k,n} = [x_{k,n,1} \ x_{k,n,2} \ x_{k,n,3} \ x_{k,n,4} \ ]^T$. The value of x_{II} is arrived at in a similar manner. Then, the value $x = x_I + x_{II}$ is the value of the combined bit allocation vector. The mutually exclusive constraint ensures that $x_{I,N}^n$ and $x_{II,N}^n$ take the shape of any of the vectors $\{[0 \ 0 \ \cdots \ 0]^T, [1 \ 0 \ \cdots \ 0]^T, [0 \ 1 \ \cdots \ 0]^T, \cdots, [0 \ 0 \ \cdots \ 1]^T\}$. This implies that just one of the components in $x_{I,N}^n$ is 1 and all others are 0s (also true for $x_{II,N}^n$). When $x_{k,n,c}$ is 1, it shows that the subchannel n has been allocated to SU k to transmit c bits per symbol. When $x_{I,N}^n$ (or $x_{II,N}^n$) has an all 0s component, it means that the subchannel n has not been allocated to any SU at all.

We define the modulation order vectors b_I and b_{II} for the two categories of SUs as follows:

$$b_I = [(b_{I,N}^1)^T \ (b_{I,N}^2)^T \ \cdots \ (b_{I,N}^N)^T]^T \ \in \mathbb{Z}^{NK_1C \times 1} \tag{8.15}$$

$$b_{II} = [(b_{II,N}^1)^T \ (b_{II,N}^2)^T \ \cdots \ (b_{II,N}^N)^T]^T \ \in \mathbb{Z}^{N(K-K_1)C \times 1} \tag{8.16}$$

where $b_{I,N}^n = [b_{I,1,n}^T \ b_{I,2,n}^T \ \cdots \ b_{I,K,n}^T]^T \ \in \mathbb{Z}^{K_1C \times 1}$ and $b_{I,k,n} = [b_{k,n,1} \ b_{k,n,2} \ \cdots \ b_{k,n,C} \ ]^T \ \in \mathbb{Z}^{C \times 1}$. The value of b_{II} is obtained in a similar fashion. Since we only considered four modulation schemes (that is, the BPSK, 4-QAM, 16-QAM and 64-QAM), $b_{1,k,n} = [1 \ 2 \ 3 \ 4]^T$ (this is also applicable to $b_{II,N}^n$).

We define the data rate matrices $B_i \in \mathbb{Z}^{K_1 \times NK_1 C}$ and $B_j \in \mathbb{Z}^{(K-K_1) \times N(K-K_1)C}$ for the two SU categories as follows:

$$B_i = \begin{bmatrix} b_1 & b_1 & \cdots & b_1 \\ b_2 & b_2 & \cdots & b_2 \\ \vdots & \vdots & \ddots & \vdots \\ b_{K_1} & b_{K_1} & \cdots & b_{K_1} \end{bmatrix}, \quad B_i \in \mathbb{Z}^{K_1 \times NK_1 C} \tag{8.17}$$

$$\begin{cases} b_1 = [b^T \ 0_C^T \ \cdots \ 0_C^T] \in \mathbb{Z}^{1 \times K_1 C} \\ b_2 = [0_C^T \ b^T \ \cdots \ 0_C^T] \in \mathbb{Z}^{1 \times K_1 C} \\ \vdots \quad \vdots \quad \ddots \quad \vdots \\ b_{K_1} = [0_C^T \ 0_C^T \ \cdots \ b^T] \in \mathbb{Z}^{1 \times K_1 C} \end{cases}$$

$$B_j = \begin{bmatrix} b_{K_1+1} & b_{K_1+1} & \cdots & b_{K_1+1} \\ b_{K_1+2} & b_{K_1+2} & \cdots & b_{K_1+2} \\ \vdots & \vdots & \ddots & \vdots \\ b_K & b_K & \cdots & b_K \end{bmatrix}, \quad B_j \in \mathbb{Z}^{(K-K_1) \times N(K-K_1)C} \tag{8.18}$$

$$\begin{cases} b_{K_1+1} = [b^T \ 0_C^T \ \cdots \ 0_C^T] \in \mathbb{Z}^{1 \times (K-K_1)C} \\ b_{K_1+2} = [0_C^T \ b^T \ \cdots \ 0_C^T] \in \mathbb{Z}^{1 \times (K-K_1)C} \\ \vdots \quad \vdots \quad \ddots \quad \vdots \\ b_K = [0_C^T \ 0_C^T \ \cdots \ b^T] \in \mathbb{Z}^{1 \times (K-K_1)C}. \end{cases}$$

From the above representations, it is now easy to write Eq. (8.4), which gives the total data rate that the network can achieve, as $\max_x [(b_I)^T x_I + (b_{II})^T x_{II}]$.

Now, we define $R_k \triangleq [R_1 \ R_2 \ \cdots \ R_{K_1}]^T \in \mathbb{R}^{K_1 \times 1}$ and $\tilde{\gamma}_k \triangleq [\tilde{\gamma}_{K_1+1} \ \tilde{\gamma}_{K_1+2} \ \cdots \ \tilde{\gamma}_K]^T \in \mathbb{R}^{(K-K_1) \times 1}$. Then, the constraint of Eq. (8.5), which gives the minimum data rate for the SUs in category one, becomes $B_i x_I \geq R_k$. Also, the constraint in Eq. (8.6), which gives the best effort data rates for the SUs in category two, becomes $B_j x_{II} = \tilde{\gamma}_k$.

We then define a power transmission vector p such that:

$$p = [(p_N^1)^T \ (p_N^2)^T \ \cdots \ (p_N^N)^T]^T \in \mathbb{R}^{NKC \times 1} \tag{8.19}$$

where $p_N^n = [p_{1,n}^T \ p_{2,n}^T \ \cdots \ p_{K,n}^T]^T \in \mathbb{R}^{KC \times 1}$ and $p_{k,n} = [p_{k,n,1} \ p_{k,n,2} \ \cdots \ p_{k,n,C}]^T \in \mathbb{R}^{C \times 1}$; $p_{k,n,c}$ is the amount of power needed to transmit c bits of data for user k on subchannel n. The power constraint in Eq. (8.7) now becomes $p^T x \leq P_{\max}$. Since the total transmission power is the sum of the transmission powers for both direct and cooperation communication, $p = p_D + p_C$, where p_D and p_C are the respective transmission power vectors for the direct and cooperation communications. The power constraint is now given as $(p_D + p_C)^T x \leq P_{\max}$.

The interference power constraints in Eqs. (8.8) and (8.9) have to be written in terms of the bit allocation vector x. To achieve this, we define a matrix $A \in \{0, 1\}^{N \times NKC}$ as follows:

$$A = \begin{bmatrix} 1_{KC}^T & 0_{KC}^T & \cdots & 0_{KC}^T \\ 0_{KC}^T & 1_{KC}^T & \cdots & 0_{KC}^T \\ \vdots & \vdots & \ddots & \vdots \\ 0_{KC}^T & 0_{KC}^T & \cdots & 1_{KC}^T \end{bmatrix}, \quad A \in \{0, 1\}^{N \times NKC} \tag{8.20}$$

$$1_{KC} = \begin{bmatrix} 1 \\ 1 \\ \vdots \\ 1 \end{bmatrix} \in \{1\}^{KC \times 1}, \qquad 0_{KC} = \begin{bmatrix} 0 \\ 0 \\ \vdots \\ 0 \end{bmatrix} \in \{0\}^{KC \times 1}.$$

If $p \odot x$ is the Schur–Hadamard (or entry-wise) product of p and x, then $A(p \odot x)$ will be that $N \times 1$ vector in which case the nth element indicates the total power used by the nth subchannel for carrying out its data transmission. We define $\varepsilon_l \triangleq [\varepsilon_1 \ \varepsilon_2 \ \dots \ \varepsilon_L]^T \in \mathbb{R}^{L \times 1}$. The constraint in Eq. (8.8), which indicates the interference power constraint for the direct communication, becomes:

$$[H_{l,n,D}^p(A(P_D \odot x))] \leq \varepsilon_l. \tag{8.21}$$

In the same vein, the constraint in Eq. (8.9), which indicates the interference power constraint for the cooperative cooperation, becomes:

$$[H_{l,n,C}^p(A(P_C \odot x))] \leq \varepsilon_l. \tag{8.22}$$

After putting all the above descriptions together, the RA problem for the heterogeneous cooperative CRN provided in Eqs. (8.4)–(8.11) can now be represented in the reformulated ILP form as follows:

$$z^* = \max_x [(b_I)^T x_I + (b_{II})^T x_{II}] \tag{8.23}$$

subject to

$$B_i x_I \geq R_k; \ k = 1, 2, \cdots, K_1 \tag{8.24}$$

$$B_j x_{II} = \tilde{\gamma}_k; \ k = K_1 + 1, K_1 + 2, \cdots, K \tag{8.25}$$

$$(p_D + p_C)^T x \leq P_{\max} \tag{8.26}$$

$$[H_{l,n,D}^p(A(p_D \odot x))] \leq \varepsilon_l \tag{8.27}$$

$$[H^p_{l,n,C}(A(p_C \odot x))] \leq \varepsilon_l \qquad (8.28)$$

$$\mathbf{0}_N \leq A\mathbf{x} \leq \mathbf{1}_N \qquad (8.29)$$

$$x_I, x_{II}, x \in \{0, 1\}. \qquad (8.30)$$

The reformulation RA problem for the heterogeneous cooperative CRN given above is now a combinatorial ILP problem. Such problems can be solved using the tool of classical optimisation. The Branch-and-Bound (BnB) method for solving ILP problems or any similarly good classical optimisation tool can be easily employed to obtain optimal (or close-to-optimal) solutions, especially when the CRN is designed as a small network. However, although such tools or methods could yield optimal solutions, the computational complexity may be significantly high, especially when the CRN is designed as a large network. It is imperative to investigate approaches that can still yield near-optimal solutions, but at a much more reduced time frame and computational demands. In this chapter, we examine the use of a heuristic for achieving such results.

8.9 Heuristic Solution for the Resource Allocation Problem in Heterogeneous Cooperative Cognitive Radio Networks

The benefits of heuristics for solving RA problems in the CRN have been well discussed in a previous chapter of this book. We now employ a fast, iterative-based heuristic to help solve the RA problem for heterogeneous cooperative CRN. While heuristics generally provide suboptimal solutions, such solutions can help in establishing the optimality-complexity trade-off, especially for large networks. Since heuristic solutions are mostly problem-specific, the particular heuristic that is developed in chapter is geared towards solving the reformulated ILP problem for heterogeneous cooperative CRN. There are two important steps involved in the heuristic, namely

- subchannel allocation and
- iterative bit and power allocation.

8.9.1 Subchannel Allocation

For the heuristic to carry out the best allocation of the subchannels to the different SU categories, it integer-relaxes the constraint $x \in [0, 1]$. This constraint now becomes:

$$0 \leq x \leq 1. \qquad (8.31)$$

What this implies is that the variable x is now permitted to be any value from 0 to 1 and is not limited to being either 0 or 1. We keep the remaining aspects of the problem formulation as they were. We then solve the new integer-relaxed problem. We obtain the values of x at the first iteration of the solution. From that solution, all the subchannels must have been assigned to the SUs that are available in the network. As a result, the data rate of SU k on the subchannel n becomes $(b_{k,n}^T x_{k,n})$.

From the initial solution obtained after relaxation of the integer constraint, it may happen that a particular SU $m \neq k$ has a higher data rate $(b_{m,n}^T x_{m,n})$ on subchannel n than the data rate that SU k has on subchannel n. The most appropriate decision would then be to give subchannel n to SU m rather than to SU k. Thus, before the subchannel n is assigned to the SU k, it must have been ascertained that $(b_{k,n}^T x_{k,n}) \geq (b_{m,n}^T x_{m,n}) \; \forall m \neq k$.

By following the guidelines for subchannel allocation given above, it becomes possible to allocate each subchannel to the SU with the possibility of achieving the best data rate on that subchannel. It must be noted that, after allocating all the subchannels to the various SUs following the guidelines at the first iteration, x reduces in dimension from the original value of $x \in [0, 1]^{KNC \times 1}$ to a lower value of $x \in [0, 1]^{NC \times 1}$.

8.9.2 Iterative Bit and Power Allocation

After allocating the subchannels to the SUs that are available in the network, what is left is to assign bits (through the modulation schemes) and transmission power to the subchannels. An iteration-based algorithm is used to achieve this. The first part of the algorithm is to allocate a conservative number of bits to each SU. The algorithm then checks the amount of power that has been used to transmit those bits, and confirms that no constraints were violated. Next, the algorithm ascertains if there is some excess power remaining. If this is true, it increases the number of bits that it assigns to the SUs in a gradual manner, where possible. Thereafter, it again reviews the amount of power that is remaining and repeats the entire iterative process if there is some power that is left. This entire process continues until it becomes impossible to further improve the bit allocation of the SUs.

Let us assume that y is the number of iterations involved in the entire process of allocating and reallocating the bits. It therefore means that, at the yth iteration step, the following optimisation problem has to be solved:

$$\max_{x^y} \; [(b_I^y)^T x_I^y + (b_{II}^y)^T x_{II}^y] \tag{8.32}$$

subject to

$$B_i x_I^y \geq [R_k - f^{(y-1)}]^+; \; k = 1, 2, \cdots, K_1 \tag{8.33}$$

$$B_j x_{II}^y = [\tilde{y}_k - g^{(y-1)}]^+; \; k = K_1 + 1, K_1 + 2, \cdots, K \tag{8.34}$$

$$(p^{(y-1)})^T x^y \leq P_{max} - \|u^{(y-1)}\|_1 \tag{8.35}$$

$$H^P[A(p^{(y-1)} \odot x^y)] \leq \varepsilon_l - H^P u^{(y-1)} \tag{8.36}$$

$$\mathbf{0}_N \leq A x^y \leq \mathbf{1}_N \tag{8.37}$$

$$\mathbf{0}_{KNC} \leq x^y \leq \mathbf{1}_{KNC} \tag{8.38}$$

where $f^{(y-1)}$ and $g^{(y-1)}$ are the respective number of bits that are allocated to the SUs in category one and category two at the yth iteration, and $u^{(y-1)}$ is the amount of power that is assigned to the SUs at the yth iteration.

We wish to further explain how the iteration process takes place. You may remember from the previous section that the allocation of bits to the subchannel n which is assigned to a SU in category one, that is, $b_{I,n} = [b_{1,n}^T \; \cdots \; b_{K_1,n}^T]^T$ is a vector of size $K_1 C \times 1$, with possible values of 1, 2, 4 and 6 in its entry. For each of representation, let us assume that there are four SUs in each category of the SUs. If we assume that, from the previous subsection, in the course of the subchannel allocation process, the first subchannel was assigned to the second SU, which is a SU in category, then, $b_{I,1} = [0\,0\,0\,0, 1\,2\,4\,6, 0\,0\,0\,0, 0\,0\,0\,0]$ for the category one SUs. If it was the third subchannel that was assigned to the first SU, which is a SU in category two, then $b_{II,3} = [1\,2\,4\,6, 0\,0\,0\,0, 0\,0\,0\,0, 0\,0\,0\,0]$ and so on.

After completing the aspect of the algorithm explained above, the algorithm identifies the elements of b_I and b_{II} that turns out to be zeros in the course of the subchannel allocation, then rename the vectors b_I and b_{II} as b_I^1 and b_{II}^1, respectively. As a result, the actual optimisation problem that is solved at the first iteration (that is, when $y = 1$) is summarised as:

$$\max_{x^1} \left[(b_I^1)^T x_I^1 + (b_{II}^1)^T x_{II}^1 \right] \tag{8.39}$$

subject to

$$B_i x_I^1 \geq R_k; \; k = 1, 2, \cdots, K_1 \tag{8.40}$$

$$B_j x_{II}^1 = \tilde{y}_k; \; k = K_1 + 1, K_1 + 2, \cdots, K \tag{8.41}$$

$$p^T x^1 \leq P_{max} \tag{8.42}$$

$$\left[H_{l,n,D}^p (A(p_D \odot x^1)) \right] \leq \varepsilon_l \tag{8.43}$$

$$\left[H_{l,n,C}^p (A(p_C \odot x^1)) \right] \leq \varepsilon_l \tag{8.44}$$

$$\mathbf{0}_N \le \mathbf{A}\mathbf{x}^1 \le \mathbf{1}_N \tag{8.45}$$

$$\mathbf{0}_{KNC,1} \le \mathbf{x}^1 \le \mathbf{1}_{KNC,1}. \tag{8.46}$$

The values of $\mathbf{f}^{(0)}$, $\mathbf{g}^{(0)}$ and $\mathbf{u}^{(0)}$ are all 0s at the first iteration, therefore, they do not reflect in the formulation above. The values of the rates $\mathbf{B}_i \mathbf{x}_I^1$ and $\mathbf{B}_j \mathbf{x}_{II}^1$, and the transmission power $\mathbf{p}^T \mathbf{x}^1$ that are obtained during the first iterative process now become the respective values $\mathbf{f}^1$, $\mathbf{g}^1$ and $\mathbf{u}^{(1)}$ for the second iterative process. The vector $\mathbf{x}^1$ is employed, alongside the power vector $\mathbf{p}$, to assign the first modulation scheme (this also implies the number of bits) for each SU on each subchannel.

The explanation earlier provided suggests that the first subchannel has been assigned to the second SU. This means that all the entries of $\mathbf{x}_I^1$ are 0s but for the elements in $x_{2,1}^1$. The algorithm calculates the total power that is assigned to the first subchannel as $(p_{2,1}^T x_{2,1}^1)$. As a generic point, the total power allocated to the SU k on the subchannel n is given as $(p_{k,n}^T x_{k,n}^1)$. We obtain the modulation scheme η (having bits c_η) that should be assigned to the SU so as not to exceed the transmission power $p_{k,n}^T x_{k,n}^1$ as:

$$\eta = arg \max_\eta \left\{ \eta \in [0, 1, 2, 3, 4] : p_{k,n,\eta} \le p_{k,n}^T x_{k,n}^1 \right\}. \tag{8.47}$$

For the purpose of clarity, what the value of η means is that it gives an idea of the highest possible modulation scheme that may be assigned to the subchannel n that will make it employ a transmission power that does not exceed the maximum power already allocated to that subchannel by the allocating algorithm. Of course, we already have a good idea on the data sizes and corresponding powers of different modulation schemes, since these values are usually finite and can be determined beforehand. As such, $p_{k,n,\eta}$ will take a set of finite power levels. By determining the bits that corresponds to the value of $p_{k,n,\eta}$, the algorithm calculates the total power that is used up to that point, the value of which will still be less than $P_{\max}$.

Since the total power that is used up is still less than $P_{\max}$, the total interference leaked to the PUs will still not be up to ε. Therefore, the likelihood of having some residual power that may still be used by the network is quite high. The excess power means that it will be possible to carry out additional iterations to help increase the number of bits that has already been allocated to the subchannels. Thus, it becomes feasible to run the second iteration ($y = 2$). Again, we note from the subchannel allocation process that the first subchannel has been allocated to the second SU, which is a category one SU, to transmit 2 bits (4-QAM modulation). Then, $\mathbf{b}_{1,2,1}$ is modified as $\mathbf{b}_{I,2,1}^2 = [0\ 0\ (4-2)\ (6-2)]^T = [0\ 0\ 2\ 4]^T$. The power $\mathbf{p}_{2,1,2}$ must have been used in the course of transmitting 2 bits on this subchannel.

If the algorithm realises that there is excess power available for use, it upgrades the allocation to maybe a 16-QAM (to transmit 2 more bits) or a 64-QAM (to transmit 4 more bits). For this action to be successful, an additional power of $(p_{2,1,3} - p_{2,1,2})$ (for the 16-QAM) or $(p_{2,1,4} - p_{2,1,2})$ (for the 64-QAM) is required. Therefore, the new power vector at the second iteration $p_{2,1}^1 =$ $[p_{2,1,1} \quad p_{2,1,2} \quad (p_{2,1,3} - p_{2,1,2}) \quad (p_{2,1,4} - p_{2,1,2})]^T$. With this, the values of the vector p^1 are realised. Assume that u_n^1 represents the power that was assigned to the subchannel n in the first iteration. Then, $u^1 \triangleq [u_1^1 \quad \cdots \quad u_N^1]^T$. This means that $P_{\max} - \sum\limits_{n=1}^{N} u_n^1$, or equivalently, $P_{\max} - \|u^1\|_1$, is now the residual power that is available for the second iteration step. By the end of the second iteration, the amount of power that has been assigned to the subchannel n is the addition of the power assigned at the first iteration and the power assigned at the second iteration. This total power is given as:

$$v_n^2 = u_n^1 + (p_{k,n}^1)^T x_{k,n}^2.$$

The newly calculated power is employed by the algorithm in deciding the new or upgraded modulation scheme η for the subchannel n.

$$\eta = arg \max_{\eta} \left\{ \eta \in [0, 1, 2, 3, 4] : p_{k,n,\eta} \leq v_n^2 \right\}. \tag{8.48}$$

Following the same explanation, we note that the total interference leaked to the PUs due to the power assigned in the first iteration step is given as $H^P u^1$. Therefore, the new permissible interference value must be less than $(\varepsilon_l - H^P u^1)$ for the second iteration. We also note that, at this second iteration, f_k^1 already represents the data rate assigned to SU k in category one during the first iteration and g_k^1 already represents the data rate assigned to SU k in category two during the first iteration. Thus, f^1 and g^1 are defined as $f^1 \triangleq [f_1^1 \quad \cdots \quad f_1^k]^T$ and $g^1 \triangleq [g_1^1 \quad \cdots \quad g_1^k]^T$, respectively.

Putting all the information together, we see that the data rate required for the SUs in category one at the second iteration would be $(R_k - f^1)$, while the available data rate for the SUs in category two at the second iteration would be $(\tilde{\gamma}_k - g^1)$. The two constraints that describe the data rate requirements now become $B_i x_i^2 \geq [R_k - f^1]^+$ for the SUs in category one and $B_j x_{II}^2 = [\tilde{\gamma}_k - g^1]^+$ for the SUs in category two.

The algorithm continues to repeat the entire iteration process. The iteration comes to a stop only when it no longer realise any significant improvement on the total achievable data rate for each SU in the system. This is indicative of the fact that a new iteration cannot improve the throughput of the system any further. The stopping criterion used to stop the iteration is as follows:

$$\left[(b_I^y)^T x_I^y + (b_{II}^y)^T x_{II}^y \right] - \left[(b_I^{y-1})^T x_I^{y-1} + (b_{II}^{y-1})^T x_{II}^{y-1} \right] = \varsigma, \tag{8.49}$$

with ς being a predetermined small value that indicates that no significant improvement in the solution can be further realised.

Table 8.1 Pseudo-code for the proposed iterative-based heuristic (Part 1)

	Pseudo-code for the subchannel allocation
1	solve for x using Eqs. (8.23)–(8.29) and (8.31)
2	set subchannel index $n = 0$
3	repeat
4	$n \leftarrow n + 1$
5	if $(b_{k,n}^T x_{k,n}) \geq (b_{m,n}^T x_{m,n}) \forall m \neq k$
6	nth subchannel is allocated to user k
7	end if
8	until $n < N + 1$

Table 8.2 Pseudo-code for the proposed iterative-based heuristic (Part 2)

	Pseudo-code for the bit and power allocation (i.e. at $y = 1, 2, 3, \ldots$)
9	set $n = 0$, $y = 0$, $u^{(0)} = 0_N$, $p^{(0)} = p$
10	repeat
11	$y \leftarrow y + 1$
12	set $f^y = 0_K$, $g^y = 0_K$, $v^y = 0_N$
13	solve the problem (8.32)–(8.38)
14	repeat
15	$N \leftarrow n + 1$
16	$v_n^y = u_n^{y-1} + (p_{k,n}^{y-1})^T x_{k,n}^y$
17	if $\eta = arg \max_\eta \left\{ \eta \in [0, 1, 2, 3, 4]: p_{k,n,\eta} \leq v_n^y \right\}$ then
18	use modulation scheme η (i.e. with c_η bits) on nth subchannel
19	set $u_{k,n}^y = p_{k,n,l}$; $f_k^y = f_k^y + c_\eta$; $g_k^y = g_k^y + c_\eta$
20	set $p_{k,n,m}^y = p_{k,n,m} - p_{k,n,l}$, $\forall m > l$
21	set $b_{k,n,m}^{y+1} = b_{k,n,l} - c_\eta$, $\forall m > l$
22	set $b_{k,n,m}^{y+1} = 0$, $\forall m \leq l$
23	end if
24	until $n < N + 1$
25	until no further improvement on total data rate (Eq. (8.49))
26	the vectors f^{y+1} and g^{y+1} contain the bits allocated for each subchannel in category one and two respectively
27	the vector u^{y+1} contains the power allocated for each subchannel

At iteration yth, the vectors $f^{(y+1)}$ and $g^{(y+1)}$ now carry the respective number of bits that has been assigned to each subchannel for the SUs in category one and category two. The vector $u^{(y+1)}$ contains the power assigned to each subchannel. The pseudo-code in Table 8.1 summarises the part of the heuristic algorithm that achieves the subchannel allocation while the Table 8.2 summarises the part of the heuristic algorithm that carries out the iterative bit and power allocation.

8.10 Useful Results from the Resource Allocation for Heterogeneous Cooperative Cognitive Radio Networks

To demonstrate the importance of the RA problem formulations and solution models for heterogeneous cooperative CRN developed in this chapter, some results are presented and discussed. The MATLAB software is used for simulating the model. The YALMIP solver is used to carry out the optimisation. The following parameters are used for the simulations: the total number of SUs in the system is 8, divided into category one SUs $K_1 = 2$, category two SUs $(K - K_1) = 2$ and the SUs from which the possible cooperator (CSU) is selected $= 4$, total number of OFDMA subchannels $N = 64$, the total number of PUs $L = 4$. The category one SUs have a minimum data rate requirement of 64 bits each. The SUs in category two use a normalised proportional rate constant γ_k, which adds to unity, to fairly share the left-over resources among them. A BER value of $\rho = 0.01$ is used for all the SUs. We chose the parameters for the simulation so that we could compare the new results with the results presented in a previous chapter, which were also similar to comparative results in the literature, such as in the results in [11] and [20].

In Figs. 8.2 and 8.3, we show the plot of the average data rates (in bits) for the different categories of SUs against the maximum interference power to the PUs for both direct and cooperative communications. In the case presented in Fig. 8.2, the maximum transmission power of the SUBS is at 20 dBm, while in the case presented in Fig. 8.3, the maximum transmission power of the SUBS is at 40 dBm. The simulation results are validated in that the results of the direct communication favourably compare with the results obtained in the works of [11, 12, 20, 21].

From the results presented in Figs. 8.2 and 8.3, one may observe that for the RA problem to have solutions that are feasible and practicable, the minimum rate requirement of the SUs in category one must always be met. The results also show

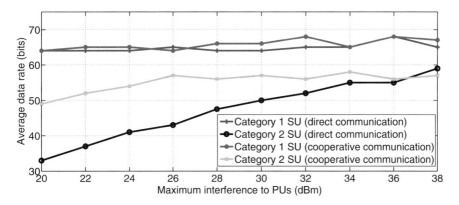

Fig. 8.2 The plot of the average data rate against the interference limit of the PUs for different SU categories. Both direct and cooperative communications are considered. The maximum SUBS power is set at 20 dBm

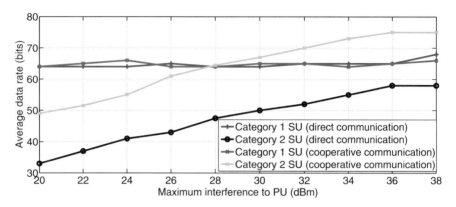

Fig. 8.3 The plot of the average data rate against the interference limit of the PUs for different SU categories. Both direct and cooperative communications are considered. The maximum SUBS power is set at 40 dBm

that, if the value of the limit of PU interference is increased, an improvement in the average data rates for the SUs in both categories is realised. However, the improvement is more pronounced for the SUs in category two than the SUs in category one. The reason is that it is easier for the allocating algorithm to gradually improve the performance of the SUs in category two once there is a slight increase in resources than it is to gradually improve the performance of the SUs in category one at a slight increase in resources. We also observe that, for all categories of SUs, the average data rate is better at a higher SUBS power (40 dBm) than at a lower SUBS power (20 dBm).

The most important observation about the results presented in Figs. 8.2 and 8.3 is the significant improvement in the performance of the network when cooperative communication is employed in comparison with employing only the direct communication. We notice that the SUs in both categories achieved a higher average data rate with cooperative communication. This is as a result of the improvement in the interference gain to PUs when cooperative communication is employed. With cooperative communication, the subchannels could transmit at a higher rate than they would have transmitted by direct communication.

We also note that in Fig. 8.2, the average data rate for the cooperative communication eventually converges to nearly that of the direct communication. This same effect would have occurred in Fig. 8.3 if we continue to increase the level of permissible interference to the PUs. A similar pattern is observed in Figs. 8.7 and 8.9, confirming this assertion. It therefore implies that as we increase the level of permissible interference to the PUs, the need for and/or effect of cooperation diminishes. Simply put, it would be more productive to use direct communication than to use cooperative communication if the PUs are well robust enough to counter the effect of the interference caused by the SUs. This is true since cooperative communication do require a lot more signalling overhead than direct communication.

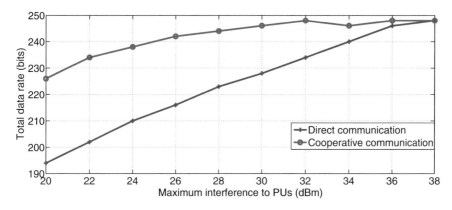

Fig. 8.4 The plot of the total data rate against the interference limit of the PUs for different SU categories. Both direct and cooperative communications are considered. The maximum SUBS power is set at 20 dBm

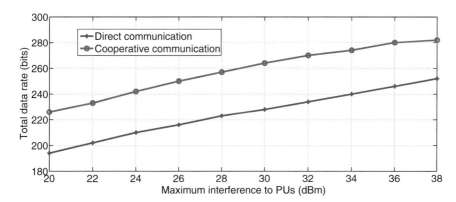

Fig. 8.5 The plot of the total data rate against the interference limit of the PUs for different SU categories. Both direct and cooperative communications are considered. The maximum SUBS power is set at 40 dBm

We plot and compare the results of the total data rate (bits) for each category of SUs against the maximum interference power to the PUs in Figs. 8.4 and 8.5. In the plots, both direct and cooperative communications are considered. Similar to the plots in Figs. 8.2 and 8.3, the maximum power of the SUBS is at 20 dBm and 40 dBm for the respective plots. Again, since the results in Figs. 8.4 and 8.5 follow the patterns of the results in Figs. 8.2 and 8.3, the prior explanations provided for Figs. 8.2 and 8.3 are all applicable to Figs. 8.4 and 8.5. Essentially, the total data rate during cooperative communication outperforms the total data rate during direct communication. The same reasoning and deductions that were made on the network performance for cooperative communication as compared with direct communication are also valid for Figs. 8.4 and 8.5.

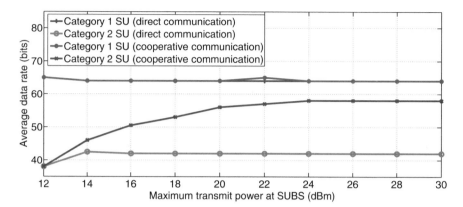

Fig. 8.6 The plot of the average data rate against an increasing SUBS power for the different SU categories. Both direct and cooperative communications are considered. The maximum interference limit of the PUs is set at 25 dBm

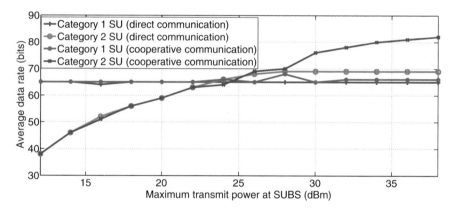

Fig. 8.7 The plot of the average data rate against an increasing SUBS power for the different SU categories. Both direct and cooperative communications are considered. The maximum interference limit of the PUs is set at 45 dBm

The plots in Figs. 8.6 and 8.7 show the results of the average data rate performance as the SUBS power is gradually increased. The results compare the performance of both direct and cooperative communications for the two categories of SUs. The maximum permissible interference limit of the PUs is at 25 dBm in Fig. 8.6, while in Fig. 8.7, the maximum permissible interference limit of the PUs is increased to 45 dBm. In the plots, only the results for the feasible regions of the RA problem are presented.

From the results presented in Figs. 8.6 and 8.7, we note that, for the problem to have feasible solutions, the minimum rate requirement of the SUs in category one has to be met at all times. Therefore, for the parameters used in the simulation, if the SUBS power is less than 12 dBm, the RA problem becomes infeasible. Another important observation is that, as we gradually increase the SUBS power, there is an improvement in the average data rate, especially for the SUs in category two. As explained previously, th reason is that since it is easier to satisfy the SUs in category two than the SUs in category one, every slight increase in resources will likely favour the category two SUs. We then note that, after a while, the data rate values peak and stabilise. Even if the SUBS power is increased, the performance does not improve any further. This is because the other constraints play their part, making it impossible for the data rate to continuously increase with every possible increase in the SUBS power.

The plots in Figs. 8.6 and 8.7 further exemplify the significant improvements that cooperative communication provides over direct communication. We note that the improvements are obvious, both when the permissible interference limit of the PUs is at 25 dBm (Fig. 8.6) and when it is at 45 dBm (Fig. 8.7). We also observe that, in Fig. 8.7, the improvements as a result of cooperation only begin to be noticeable at a SUBS power of about 26 dBm. What this implies is that the network prefers to use direct communication when the SUBS has limited power in order to maximise the use of its transmission power, and to reduce signalling overhead. As the SUBS power increases though, the network prefers to use cooperative communication which results in a better performance for the network.

It is also noteworthy that in Fig. 8.7 (a similar pattern is observed in Fig. 8.9), the value of the average data rate (and the total data rate, as shown in Fig. 8.9) during cooperative communication eventually saturates despite the continuous increase of the maximum transmission power of the SUBS. The improvements in the average and total data rates are not indefinite because the other constraints also affect the network, eventually limiting the possible improvement in performance that the network can realise.

The plots in Figs. 8.8 and 8.9 are the results for the total data rate plotted against an increasing power at the SUBS. The plots compare the results for both direct and

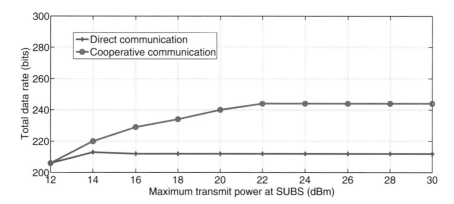

Fig. 8.8 The plot of the total data rate against an increasing SUBS power for the different SU categories. Both direct and cooperative communications are considered. The maximum interference limit of the PUs is set at 25 dBm

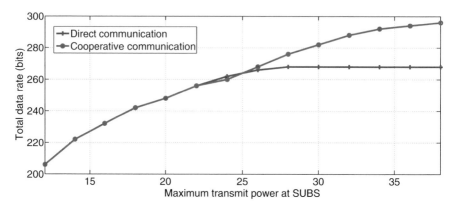

Fig. 8.9 The plot of the total data rate against an increasing SUBS power for the different SU categories. Both direct and cooperative communications are considered. The maximum interference limit of the PUs is set at 45 dBm

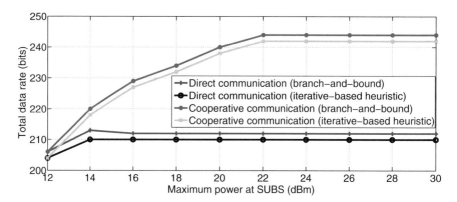

Fig. 8.10 Comparing the performance of the ILP solution to the heuristic solution. The comparison is based on the total data rates. The permissible interference limit of the PUs is set at 25 dBm

cooperative communications for all the SU categories. In Fig. 8.8, the permissible interference limit for the PUs is set at 25 dBm, while in Fig. 8.9, the permissible interference limit for the PUs is set at 45 dBm. The explanations given for Figs. 8.6 and 8.7 are equally applicable to Figs. 8.8 and 8.9. Essentially, the performance of the total data rates during cooperative communication is better than during direct communication. The reasons are the same as given for the results in Figs. 8.6 and 8.7.

The plots in Figs. 8.10 and 8.11 are comparative results obtained for the ILP solution (using the BnB approach) and the heuristic solution (using the iterative algorithm) for the heterogeneous cooperative CRN discussed in this chapter. The results compare the optimality and the computational complexities of the solutions. For the ILP solution, the computational complexities are obtained from the number

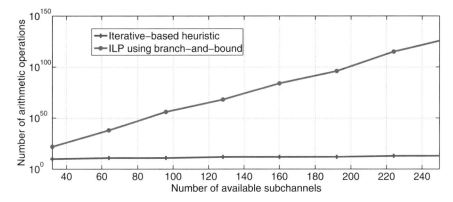

Fig. 8.11 Comparing the performance of the ILP solution to the heuristic solution. The comparison is based on the computational complexity for different number of subchannels

of arithmetic operations needed to arrive at the solution [21]. For the heuristic solution, the total complexity is obtained by summing the complexities of the two parts of the algorithm (that is, the subchannel allocation and the iterative bit and power allocation). The results in Figs. 8.10 and 8.11 indicate that the performance of the heuristic, in terms of the total data rates, is fairly close to the performance of the ILP technique. However, the complexity demand for the heuristic solution is much less than for the ILP solution, especially as the CRN becomes larger.

A good inference for network designers can be drawn from the results in Figs. 8.10 and 8.11. The inference is that, for practical CRN implementation, once a good idea on the optimal results has been established, it is advisable to develop heuristic solutions that are sufficiently close to the optimal solutions to help solve the RA problems. This will ensure that the solutions provided are feasible, timeous and computationally less demanding, especially for large CRN considerations.

8.11 Summary of the Chapter

This chapter has established the importance of developing RA problem formulations and solution models that can achieve outstanding performances, despite the stringent interference constraints imposed on the CRN. In the chapter, the concept of cooperative diversity was employed to help mitigate the limiting effects of interference to the PUs of the CRN. This resulted in significant improvements in the RA solutions for the heterogeneous cooperative CRN. Thus, appropriate cooperative diversity methods can be incorporated to modern CRN designs in order to mitigate the effects of interference and to achieve optimal results. For practical, realistic CRN designs, heuristics and other computationally less demanding optimisation tools may be used to solve the RA problems for heterogeneous cooperative CRN at a much reduced time duration and complexity. The performance improvements in the RA solutions for the CRN when cooperation is incorporated are quite remarkable, as the results presented in this chapter clearly show.

References

1. L.E. Doyle, *Essentials of Cognitive Radio*. The Cambridge Wireless Essentials Series (Cambridge University Press, New York, 2009)
2. K. Pretz, Overcoming spectrum scarcity – cognitive radio networks might be one answer (2012). http://theinstitute.ieee.org/technology-focus/technology-topic/overcoming-spectrum-scarcity
3. M. Ge, S. Wang, On the resource allocation for multi-relay cognitive radio systems, in *Proceedings of the IEEE ICC* (2014), pp. 1591–1595
4. J. Li, T. Luo, G. Yue, Resource allocation scheme based on weighted power control in cognitive radio systems, in *Proceedings of the ICCCAS*, vol. 1 (2013), pp. 178–182
5. B.S. Awoyemi, B.T. Maharaj, Mitigating interference in the resource optimisation for heterogeneous cognitive radio networks, in *Proceedings of the IEEE 2nd Wireless Africa Conference (WAC)* (2019), pp. 1–6
6. J. Oh, W. Choi, A hybrid cognitive radio system: a combination of underlay and overlay approaches, in *Proceedings of the IEEE VTC (Fall)* (2010), pp. 1–5
7. N. Hao, S.-J. Yoo, Interference avoidance throughput optimization in cognitive radio ad hoc networks. EURASIP J. Wirel. Commun. Netw. **2012**(1) (2012). http://dx.doi.org/10.1186/1687-1499-2012-295
8. Z. Chen, C.-X. Wang, X. Hong, J. Thompson, S. A. Vorobyov, F. Zhao, X. Ge, Interference mitigation for cognitive radio {MIMO} systems based on practical precoding. Phys. Commun. **9**, 308–315 (2013). http://www.sciencedirect.com/science/article/pii/S1874490712000390
9. D. Hu, S. Mao, On co-channel and adjacent channel interference mitigation in cognitive radio networks. Ad Hoc Netw. **11**(5), 1629–1640 (2013). http://www.sciencedirect.com/science/article/pii/S1570870513000292
10. B.S. Awoyemi, B.T.J. Maharaj, A.S. Alfa, Solving resource allocation problems in cognitive radio networks: a survey. EURASIP J. Wirel. Commun. Netw. **2016**(1), 176 (2016). https://doi.org/10.1186/s13638-016-0673-6
11. B.S. Awoyemi, B.T. Maharaj, A.S. Alfa, Resource allocation for heterogeneous cognitive radio networks, in *Proceedings of the IEEE WCNC* (2015), pp. 1759–1763
12. B.S. Awoyemi, B.T. Maharaj, A.S. Alfa, QoS provisioning in heterogeneous cognitive radio networks through dynamic resource allocation, in *Proceedings of the IEEE AFRICON* (2015), pp. 1–6
13. B. Awoyemi, T. Walingo, F. Takawira, Predictive relay-selection cooperative diversity in land mobile satellite systems. Int. J. Satellite Commun. Netw. **34**(2), 277–294 (2016). https://doi.org/10.1002/sat.1118
14. B. Awoyemi, T. Walingo, F. Takawira, Relay selection cooperative diversity in land mobile satellite systems, in *Proceedings of the IEEE AFRICON* (2013), pp. 1–6
15. J. Laneman, D. Tse, G. W. Wornell, Cooperative diversity in wireless networks: efficient protocols and outage behavior. IEEE Trans. Inf. Theory **50**(12), 3062–3080 (2004)
16. S. Wang, M. Ge, C. Wang, Efficient resource allocation for cognitive radio networks with cooperative relays. IEEE J. Sel. Areas Commun. **31**(11), 2432–2441 (2013)
17. M. Adian, H. Aghaeinia, Optimal resource allocation for opportunistic spectrum access in multiple-input multiple-output-orthogonal frequency division multiplexing based cooperative cognitive radio networks. IET Signal Process. **7**(7), 549–557 (2013)
18. S. Du, F. Huang, S. Wang, Power allocation for orthogonal frequency division multiplexing-based cognitive radio networks with cooperative relays. IET Commun. **8**(6), 921–929 (2014)
19. M. Pischella, D. Le Ruyet, Cooperative allocation for underlay cognitive radio systems, in *Proceedings of the 14th IEEE Workshop on SPAWC* (2013), pp. 245–249

20. Y. Rahulamathavan, K. Cumanan, L. Musavian, S. Lambotharan, Optimal subcarrier and bit allocation techniques for cognitive radio networks using integer linear programming, in *Proceedings of the 15th IEEE Workshop on SSP* (2009), pp. 293–296
21. Y. Rahulamathavan, S. Lambotharan, C. Toker, A. Gershman, Suboptimal recursive optimisation framework for adaptive resource allocation in spectrum-sharing networks. IET Signal Process. **6**(1), 27–33 (2012)

Chapter 9
Interference Management and Control in Cognitive Radio Networks Using Stochastic Geometry

9.1 Interference Management Through Stochastic Geometry

In most modern wireless communication networks, the problem of interference is one of the greatest challenges that must be overcome when allocating network resources. Interference has the tendency of degrading the performance of any wireless communication network if not properly controlled. Especially in the CRN, the problem of interference is even more exacerbated since the SUs have to transmit on the scarce spectrum resources belonging to the PUs. In the CRN, the SUs aim to satisfy their spectrum demands, while still ensuring that their transmissions do not generate excessive interference in the primary networks [1]. When interference is not properly managed and controlled in the CRN, the essence of the CRN is defeated since channel usage can neither be effective nor efficient. Interference management and control therefore remain an important aspect of the CRN and must be properly characterised to ensure an efficient resource allocation process. Recently, the tool of stochastic geometry (SG) is being employed as an important tool for achieving interference management and control in modern CRN.

To help understand the concept of SG as a tool for interference management, the CRN can be considered as a collection of transmitting and receiving nodes, located within a certain domain with different priority levels (generally classified into primary and secondary priority levels). In the CRN, multiple primary and secondary transmitter–receiver pairs transmit on the scarce spectrum resources, therefore, the intended signal at any receiver can be interfered with by the transmissions of other transmitting nodes. It is important to note that, in such a network, the pattern of users' distributions is an important element of the system and must be properly considered when investigating the signal-to-interference plus noise ratio (SINR) or signal-to-inference ratio (SIR) in an interference-limited network, at any primary or secondary receiver with the aim of reducing interference in the network.

In order to capture these patterns of users' distributions, the adoption of the hexagonal grid model was earlier considered in wireless communications. Such a

© The Author(s), under exclusive license to Springer Nature Switzerland AG 2022
B. TJ Maharaj, B. S. Awoyemi, *Developments in Cognitive Radio Networks*,
https://doi.org/10.1007/978-3-030-64653-0_9

model has been proven to be unsuitable, especially when the actual distributions of users in practical networks and other unique properties of networks are taken into consideration. On the other hand, SG is known to provide a useful way of obtaining these macroscopic properties of wireless communication networks through the use of point processes—the element often considered as the most important aspects of SG. The SG is a branch of mathematical research that provides an opportunity to study the random phenomena of nodes within any dimensional area. Through the tool of SG, the spatial distributions of users/nodes can be properly realised and accurately evaluated.

The adoption of SG now continues to receive attention in the CRN. With this tool, the distributions of PUs and SUs in a typical CRN can be captured in order to model interference within such networks. Interference modelling is important towards ensuring that the transmissions of PUs are not disrupted as a result of the channel access opportunities given to the SUs. When modelling the distributions of users in the CRN, two properties of the CRN are significant to the performance of the network. These properties are as follows:

- *The dependent distribution property*: This is the property that shows that the distributions of PUs and SUs are closely dependent.
- *The priority property*: This is the property that represents the pre-emptive priority of PUs over SUs.

These important characteristics are now being considered when modelling interference in CRN using SG.

One of the benefits of SG, when adopted in the CRN, is its capability to provide solutions that are not only accurate but are also scalable, especially when multiple nodes are considered as in the large-scale CRN. Although most authors often adopt various simplification techniques in order to obtain tractable analyses for some metrics of interests, SG can provide accurate analyses when these simplifications are carefully achieved. Its ability to provide tractable and accurate solutions has been demonstrated and verified by many existing works in various wireless communications networks. Its applications in the CRN can enhance interference management and control, thereby ensuring an efficient and effective resource allocation process among various users. As a result, the SG approach has gained more interest over the conventional grid model in recent years.

9.2 Advantages of Stochastic Geometry over the Conventional Hexagonal Grid Model

When the distributions of users are represented following the hexagonal grid model, the locations of transmitters are generally considered to be at the centre of hexagonal lattices where the channels are regularly spaced such that there is no intra-channel interference. Although such a technique can reduce interference when used, its limitations mean that the grid-based models are not very suitable to cope with the dynamics required for xG networks, such as the CRN. The advantages of SG-based techniques over the grid-based models are summarised as follows:

- **Accuracy:** When characterising interference in any wireless network, accuracy is the most important requirement of any interference model. An interference model that overestimates interference will deny SUs opportunities to access the spectrum resources even when there are spectrum opportunities, while an interference model that underestimates interference in the network will degrade the performance of the network. The grid-based models have been demonstrated to be inaccurate, owing to the patterns of users' distributions, as well as its unrealistic assumptions [2]. Hence, grid-based models are unsuitable for CRN applications. In contrast, the SG-based technique provides tightly bound solutions when used and its accuracy has been demonstrated in many works, for example, the works in [3–5]. These SG-based solutions have been shown to be reasonably accurate when compared with real-life data [2, 4, 6].
- **Tractability:** Tractability is another important requirement of any interference model. Analyses obtained through the hexagonal grid models are not tractable and, in fact, they require complex Monte Carlo simulations [7]. Other research studies have considered the use of the Wyner model due to its tractability. Such models are, however, impractical while the results are generally inaccurate [4]. On the other hand, SG tools are able to provide analyses that are tractable. With such an approach, closed-form, as well as tightly bound and approximate expressions can be obtained for various metrics of interests.
- **Scalability:** As a result of the idealised nature of the grid-based modelling approach when representing the distributions of users, its solutions are not scalable, especially when multiples users are involved. Grid-based models are not capable of improving the channel usage efficiency because of the inability to characterise the locations of users in more realistic terms. Interestingly, SG tools are known for their ability to realistically model users' locations using stochastic point processes. With these point processes, scalable solutions can be obtained. Hence, coverage, rate and mobile users' experience in networks with SG-based models are quite different than in networks with traditional grid-based models.

9.3 Users' Distributions Modelling in Cognitive Radio Networks

As previously mentioned, the limitations of the grid-based models are mainly due to their oversimplification of users' distributions over d-dimensional space. On the other hand, SG-based models through the point processes allow the statistical properties of a random collection of nodes to be properly described and derived within the d-dimensional space. These point processes are generally defined as a finite random collection of points within a measured space and are significant when characterising users' spatial locations in CRN. A point process with a constant intensity of points within the Euclidean space R^d is known to be homogeneous and non-homogeneous if otherwise. Similarly, point processes are often described as stationary, simple and isotropic. A stationary point process is usually considered to be invariant by translation, while a point process is simple (that is, $N(\{x\}) \leq 1, \forall x \in R^d$) if no two points exist in the same location. Finally, a point process that is invariant to motion is isotropic.

In order to properly represent the relationships between the distributions of PUs and SUs, various point processes can be used to model users' locations while producing tractable and accurate analyses. These point processes have been extensively discussed in the literature, while their suitability has also been established. The common ones when modelling interference in the CRN are the Poisson point process (PPP), the hardcore process (such as the Matern hardcore point process (MHCP), the Poisson hole process (PHP)) and the Poisson cluster process (PCP). The binomial point process (BPP) is another point process but has not been well considered for the CRN. These point processes are discussed next for more insights.

9.3.1   Poisson Point Process

Often described as the most important point process, the PPP is the most used and tractable point process among all presently known point processes [8]. It is therefore unsurprising that most of the existing works have focussed on its adoption when modelling interference in the CRN. A PPP can be homogeneous, non-homogeneous, stationary, simple or isotropic. Considering a stationary PPP Ψ of intensity μ, the number of points n within a bounded set $B \subset R^2$ is a Poisson distribution with mean $\mu|B|$ and can be expressed as:

$$P[\Psi(B) = n] = \frac{(\mu|B|)^n}{n!} \exp^{-\mu|B|}. \qquad (9.1)$$

Note that for all disjoint set $B_i \subset R^2 (\forall i = 1, 2, \ldots, k)$, $(\Psi(B_i), \ldots, \Psi(B_k))$ are independent. This implies that all users are assumed to be identically and independently distributed within the region under consideration when users' distributions are modelled using PPP. For instance, when the distributions of PUs and SUs are modelled using two independent homogeneous PPP, users' distributions are said to be independent, and hence, two or more users can be arbitrarily close to each other than possible in practical systems. The realisation of users' distributions under the PPP assumption is presented in Fig. 9.1.

Modelling users' spatial distributions in the CRN using independent homogeneous PPP is not only analytically tractable but it is also known to be capable of producing remarkable analytical solutions for important metrics such as coverage and outage probability, medium access probability, spectral efficiency, etc. Despite this, a quick observation of Fig. 9.1 shows that the PUs and SUs' distributions in

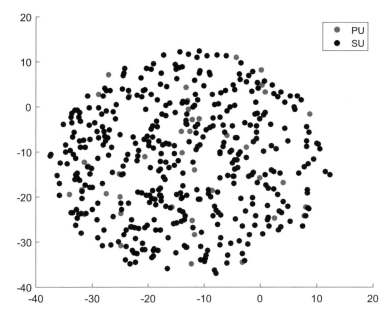

Fig. 9.1 The realisation of users' distributions under the PPP assumption

real life may not actually follow two independent PPPs since the SUs are expected to ensure a minimum distance to active PUs, so as to avoid interfering with the PUs' transmissions. In fact, practical deployments of PUs do not follow PPP because of the fact that a minimum repulsion is often observed among the PUs (for example, in TV transmitters) in practical systems. As a result, the need to investigate better point processes becomes a necessity. Nevertheless, the application of PPPs continues to receive wide adoption because of its tractability. Authors such as [9] sometimes adopt the concept of void probability in order to capture the expected repulsions among active transmitting nodes.

9.3.2 Binomial Point Process

Although homogeneous PPP is analytically convenient and often produces tractable analyses, its usage may not be appropriate in networks with a known number of users where a fixed and finite number of interfering nodes are distributed randomly within the area of interest. Also, users with locations closer to the centre of the network may experience more interference than users that are closer to the boundary of the network [10]. In order to find a more useful point process when modelling interference in fixed and finite networks, the BPP can be considered. Its application has been limited since the probability density function for interference when BPP is considered is unknown [10].

The realisation of BPP is similar to the realisation of PPP presented in Fig. 9.1 except for the known number of points in BPP. Considering a scenario where the distribution of users within a bounded region B follows BPP Ψ_{BPP}, such a distribution is a superposition of n independent and uniformly distributed points in B. The probability that the number of nodes within $C, \forall C \subset B$ is m can be obtained as:

$$P(\Psi_{BPP}(C) = m) = \binom{n}{m}\left(\frac{|C|}{|B|}\right)^{m}\left(1 - \frac{|C|}{|B|}\right)^{n-m}. \tag{9.2}$$

9.3.3   Hardcore Point Processes

Under the PPP and BPP, the distributions of PUs and SUs were assumed to be identical and independent. Such an assumption neglects the minimum repulsions required among PUs and SUs. Consider a CRN where PUs are deployed in such a way that there is a minimum repulsion of radius D within any two PUs. This region of radius D belonging to any typical PU is known as the exclusion region of such a PU within which no SU can be active. This technique of exclusion regions is capable of reducing interference among PUs, as well as interference from active SUs at the primary network.

When exclusion regions are introduced in the primary networks, the distributions of PUs are better depicted as MHCP [10]. This is because of the capability of MCHP to capture the required repulsions among PUs. The distribution of users in the CRN when MHCP is adopted in the primary network is presented in Fig. 9.2. This is achieved through a dependent thinning of Fig. 9.1 to obtain the required repulsions among users in the network. Despite its capability to model users' distributions closer to practical networks, there is no known probability generating functional (PGFL) for MHCP [8, 10]. Hence, only the approximate form of its function can be obtained [8].

Despite the non-availability of its PGFL, the MHCP assumption is a very useful approach towards interference management and control in the CRN. With such an approach, the intra-network interference in the primary network is significantly reduced while the interference generated through the activities of the SUs at the active PUs is reduced due to the minimum distance between any active PU and the nearest active SU under such an assumption. Similarly, the hardcore processes such as PHP and MHCP can be used in the secondary networks. When used, active SUs are also separated by a minimum repulsion such that a typical SU is only allowed to transmit if its location is not within the locations of all the active PUs and the currently active SUs. The adoption of hardcore point processes is, however, limited in the CRN because of the difficulty in obtaining closed-form expressions for various metrics when used.

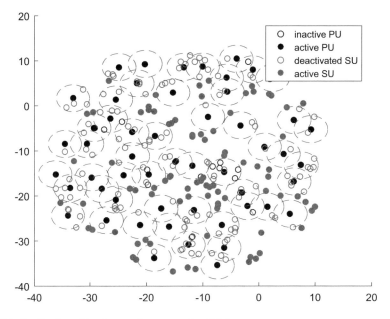

Fig. 9.2 Realisation of hardcore point process for CRN

There are two popular methods for the detection of the exclusion region (which is also known as the protection region). These methods are distance-based and threshold-based detection mechanisms.

9.3.3.1 Distance-Based Exclusion Regions

In the distance-based exclusion regions, the PUs' protection zones are determined by the SUs through estimation of the distance between a typical secondary transmitter (ST) and the test primary receiver (PR) under the receiver-centric model or the distance between a typical ST and serving primary transmitter (PT) under the transmitter-centric model. Under the transmitter-centric model, SUs can estimate the distance through the PT signal power received at the typical ST. In the receiver-centric model, the distance between the tagged PR and the typical ST is determined through beacons that are generated by active receivers and transmitted via a dedicated channel.

Considering the CRN presented Fig. 9.3, if a test PT with a transmission power P_p is located at a distance $r_{p,o}$ from its paired PR, the nearest PT asides the test PT will be located at a distance $r_{p,i} > r_{p,o}$ from the tagged PR in a receiver-centric model when assuming that a receiver is connected to the nearest transmitter. In order to ensure that the interference at the tagged PR is within the acceptable limit, the nearest ST with transmission power P_s will be located at the distance r_s given as

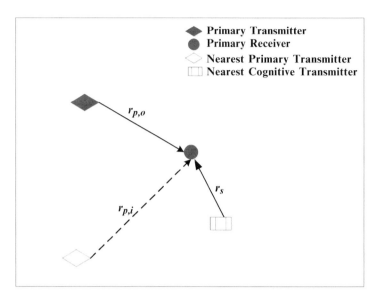

Fig. 9.3 The distance-based exclusion region

$r_s > r_{p,o} \left(\frac{P_s}{P_p} \right)^{\left(\frac{1}{\eta} \right)}$, provided that both networks have the same path loss exponent η [8].

9.3.3.2   Threshold-Based Exclusion Regions

Unlike in the distance-based exclusion region where any ST determines the PUs' exclusion regions through distance, STs rely on their sensed threshold to determine the exclusion regions under the threshold-based exclusion region. This concept was used for opportunistic spectrum access in CRNs [11]. In this approach, a typical ST is located within the exclusion region of a PU if the power received at the SU is greater than the predefined threshold for primary transmissions. This approach can also be adopted in any secondary network in which case a typical ST's location will be considered to be within the exclusion region of another SU if the power received is greater than the predefined threshold for secondary transmissions. The predefined thresholds for both primary and secondary transmissions have to be carefully estimated for interference control in the network. Similar to a distance-based approach, a threshold-based approach can be either transmitter-centric or receiver-centric, depending on whether the signal received is from the PT or PR.

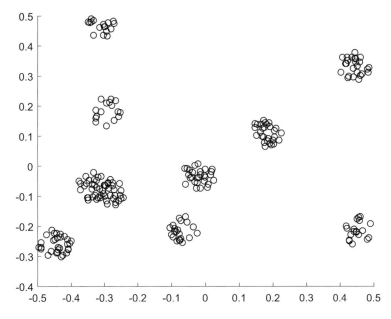

Fig. 9.4 The realisation of PCP

9.3.4   Poisson Cluster Process

The last point process discussed in this chapter is the PCP. Sometimes, the locations of users can follow a clustering pattern, for instance, when the presence of SUs is restricted to a certain part of the network. In such a network, users' distributions are better captured using PCP. Cluster-based point processes are difficult to analyse and may lead to analyses that are non-tractable. Other examples of cluster-based point processes are the Thomas cluster process and the Matern cluster process. For a clearer understanding of the process, the realisation of the PCP is shown in Fig. 9.4.

9.4   Analysis of the Signal-to-Interference Plus Noise Ratio

The SINR and SIR are very powerful tools when modelling interference in any wireless network. In a network where the effect of noise is non-significant, the analysis for SIR is generally obtained with noise signal power neglected. Note that under the receiver-centric network, each PT is considered to be located within the region centred on its respective corresponding PR. Similarly, each ST is located within the region centred on its paired secondary receiver (SR). Under the transmitter-centric scenario, transmitters are located at the centre of their respective region with each receiver uniformly located within these regions.

Consider a typical PR Y_o located at the origin of a disk. The SINR at such a PR in the two- or three-dimensional Euclidean space can be expressed as:

$$SINR_{Y_o} = \frac{S_p}{W + I_{pp} + I_{sp}}, \tag{9.3}$$

where S_p is the desired signal power from the tagged PT X_o, W is the noise signal power, I_{pp} is the interference received from all PTs except the tagged PT and I_{sp} is the interference received from STs. Similarly, consider a typical SR y_o located at the centre of a disk with its paired ST x_o located within such a disk. The SINR at such an SR is also obtained as:

$$SINR_{y_o} = \frac{S_s}{W + I_{ss} + I_{ps}}, \tag{9.4}$$

where S_s is the desired signal power from the tagged ST x_o, I_{ss} is the interference received from all STs except the tagged ST and I_{ps} is the interference received from PTs. The desired signal power S_p is given as:

$$S_p = P_p h_{X_{pp}} ||X_{pp}||^{-\eta}, \tag{9.5}$$

where P_p is the transmit power of the tagged PT, $h_{X_{pp}}$ is the random channel gain that captures the outcome of fading and shadowing between any primary transmitter–receiver pair, $||X_{pp}||$ is the Euclidean distance between any tagged primary transmitter–receiver pair and η is the path loss exponent. In the secondary network, the desired signal power S_s is given as:

$$S_s = P_s h_{x_{ss}} ||x_{ss}||^{-\eta}, \tag{9.6}$$

where P_s is the transmission power of the ST, $h_{x_{ss}}$ is the random channel gain that captures the outcome of fading and shadowing between any secondary transmitter–receiver pair and $||x_{ss}||$ is the Euclidean distance between any tagged secondary transmitter–receiver pair. The expressions in Eqs. (9.3)–(9.6) are central to interference modelling in the CRN.

9.4.1 Interference Modelling in the Primary Network

The purpose of interference management and control in CRN is to ensure coverage in the primary network while meeting the channel access requirements of the SUs. In order to achieve this, analyses are often obtained for several performance metrics based on the analyses of the SINR presented in Eqs. (9.3) and (9.4). The most used metrics are the *probability of successful transmission* (which is analytically the same as coverage probability), *outage probability, spectral efficiency, medium access probability*, etc. The analyses for the first two metrics are closely related and follow similar techniques. Without introducing any unnecessary complexity, a transmission between any test transmitter–receiver pair can be said to be successful if the SINR received at the test receiver is greater than the predefined SINR threshold θ. The outage probability is the complement of the probability of successful transmission and is equivalent to the cumulative distribution function of the SINR [12].

At any tagged PR Y_o, the probability that the packet sent by its paired PT is successfully received is given as:

$$P_{suc}(Y_o) = P(SINR_{Y_o} > \theta_p), \tag{9.7}$$

where θ_p is the predefined SINR threshold for primary transmissions. From Eqs. (9.3) and (9.5), it is clear that:

$$
\begin{aligned}
P_{suc}(Y_o) &= P\left(\frac{P_p h_{X_{pp}} ||X_{pp}||^{-\eta}}{W + I_{pp} + I_{sp}} > \theta_p\right), \\
&= P\left\{h_{X_{pp}} > \frac{\theta_p(W + I_{pp} + I_{sp})}{P_p ||X_{pp}||^{-\eta}}\right\}, \\
&= P\left\{h_{X_{pp}} > \left(\frac{\theta_p ||X_{pp}||^{\eta}}{P_p}W + \frac{\theta_p ||X_{pp}||^{\eta}}{P_p}I_{pp} + \frac{\theta_p ||X_{pp}||^{\eta}}{P_p}I_{sp}\right)\right\}.
\end{aligned}
\tag{9.8}
$$

Because of the necessity of tractable analysis and exact distribution for the SINR, a common assumption is that Rayleigh fading is experienced between the test PR and its corresponding serving PT [8]. With the assumption of Rayleigh fading, the channel power gain $h_{X_{pp}} \sim \exp(1)$. Without loss of generality, the Euclidean distance between the test primary transmitter–receiver pair is generally considered to be constant, that is, $r_p = ||X_{pp}||$. From this, we have:

$$P_{suc}(Y_o) = E\left\{\exp\left(\frac{-\theta_p ||X_{pp}||^{\eta}}{P_p}W - \frac{\theta_p ||X_{pp}||^{\eta}}{P_p}I_{pp} - \frac{\theta_p ||X_{pp}||^{\eta}}{P_p}I_{sp}\right)\right\}. \tag{9.9}$$

Note that the Laplace Transform (LT) of z is given as $L_z(s) = E[\exp(-sz)]$. From this definition, it is clear that Eq. (9.9) can be obtained at $s = \frac{\theta_p ||X_{pp}||^{\eta}}{P_p}$ as:

$$P_{suc}(Y_o) = \exp(-sW)L_{I_{pp}}(s)L_{I_{sp}}(s), \tag{9.10}$$

where $L_{I_{pp}}$ and $L_{I_{sp}}$ are the LTs of I_{pp} and I_{sp}, respectively. Obtaining the solutions for the LTs can be difficult and usually depends on the distribution assumed to realise the locations of users in the network. Few cases are considered to provide some insights.

Case 1 Consider a typical CRN where the PTs are distributed following a homogeneous PPP Ψ_p of intensity μ_p and the STs are distributed following an independent PPP Ψ_s of intensity μ_s. The derivation for I_{pp} can be obtained as interference from all active PTs except the tagged PT and is given as:

$$I_{pp} = \sum_{X_i \in \Psi_p \backslash X_o} P_p h_{X_{pp}} ||X_{pp}||^{-\eta}. \tag{9.11}$$

Its LT in Eq. (9.10) is then given as:

$$\mathcal{L}_{I_{pp}}(s) = E\left[\exp\left(-s \sum_{X_i \in \Psi_p \backslash X_o} P_p h_{X_{pp}} ||X_{pp}||^{-\eta} \right) \right],$$

$$= E\left[\prod_{X_i \in \Psi_p \backslash X_o} \exp(-s P_p h_{X_{pp}} ||X_{pp}||^{-\eta}) \right]. \tag{9.12}$$

From the PGFL of PPP, we know that:

$$E\left[\prod_{X_i \in \Psi_p \backslash X_o} f(x) \right] = \exp\left(-\mu_p \int_{R^2} (1 - f(x)) dx \right). \tag{9.13}$$

Hence,

$$\mathcal{L}_{I_{pp}}(s) = \exp\left(-\mu_p \int_{R^2} (1 - \exp(-s P_p h_{X_{pp}} ||X_{pp}||^{-\eta})) dX \right). \tag{9.14}$$

With the assumption of Rayleigh fading, $h_{X_{pp}} \sim \exp(1)$. As a result:

$$\mathcal{L}_{I_{pp}}(s) = \exp\left(-\mu_p \int_{R^2} (1 - \exp(-s P_p ||X_{pp}||^{-\eta})) dX \right),$$

$$= \exp\left(-\mu_p \int_{R^2} 1 - \exp\left(\frac{-s P_p}{||X_{pp}||^{\eta}} \right) dX \right),$$

$$\mathcal{L}_{I_{pp}}(s) = \exp\left(-\mu_p \int_{R^2} \frac{1}{1 + \frac{||X_{pp}||^{\eta}}{s P_p}} dX \right). \tag{9.15}$$

Following some algebraic manipulations, Eq. (9.15) is converted from the Cartesian coordinate to the polar coordinate form so as to obtain a closed-form expression for $\mathcal{L}_{I_{pp}}$. At $\tau = \frac{2}{\eta}$,

$$\mathcal{L}_{I_{pp}}(s) = \exp\left[-\pi \mu_p \frac{(s P_p)^\tau}{sinc(\tau)} \right], \tag{9.16}$$

$$\mathcal{L}_{I_{pp}}(s) = \mathcal{L}(\mu_p, P_p, s). \tag{9.17}$$

Similarly, the expression for $\mathcal{L}_{I_{sp}}$ is obtained following $I_{sp} = \sum_{x_i \in \Psi_s} P_s h_{x_{sp}}$ $\|x_{sp}\|^{-\eta}$. From [13]:

$$\mathcal{L}_{I_{sp}} = \exp\left(-2\pi\mu_s^a \int_D^\infty \frac{r}{1 + \frac{r^\eta}{sP_s}}\right), \tag{9.18}$$

$$\mathcal{L}_{I_{sp}}(s) = \mathcal{L}(\mu_s^a, P_s, s, D). \tag{9.19}$$

The exact analysis for $\mathcal{L}_{I_{sp}}$ is difficult to obtain and the one provided in Eq. (9.18) is only an approximate expression for it [12]. Note that the exclusion region of radius D is implemented around each PR. Hence, I_{sp} is dominated by interference from STs outside the exclusion region centred on the test PR [12]. The intensity μ_s^a is obtained following an independent thinning of the SUs outside the exclusion regions with probability $\exp(\mu_p\pi D^2)$ using the void probability technique. It is worth noting that when only the STs outside the exclusion regions of the PTs are considered, the distribution of the active STs is no longer a PPP but a PHP. The analysis in Eq. (9.18) has, however, been shown to be a good approximation of I_{sp} [13]. The upper bound for the $\mathcal{L}_{I_{sp}}$ can be expressed as:

$$\mathcal{L}_{I_{sp}}^{upper}(s) \triangleq \mathcal{L}(\mu_s, P_s, s, D). \tag{9.20}$$

Case 2 Consider a typical CRN where the PTs are distributed following a HPPP Ψ_p^a of intensity $\mu_p^a = \mu_p \exp(-\mu_p\pi D^2)$, while the active STs are distributed following a PHP Ψ_p^{php} of intensity $-\mu_s^{php}$. Similarly, each serving PT is assumed to be located within the exclusion region of radius D centred on its paired PR, while each secondary transmitter–receiver pair is separated by a distance r_s following the bipolar network model (details for obtaining the parameter D through the bipolar network model are provided in [12]). The definition of $\mathcal{L}_{I_{pp}}$ can be given as:

$$\mathcal{L}_{I_{pp}}(s) \triangleq \mathcal{L}(\mu_p^a, P_p, s), \tag{9.21}$$

while the definition for $\mathcal{L}_{I_{sp}}$ can be approximated as:

$$\mathcal{L}_{I_{sp}}(s) \triangleq \mathcal{L}(\mu_s^{php}, P_s, s), \tag{9.22}$$

where $\mu_s^{php} = \mu_s \exp(-\mu_p\pi D^2)$.

As previously stated, the analysis for PHP is non-tractable and its PGFL is unknown hence the need for the approximation presented in Eq. (9.22) [12, 14]. In [15], the authors showed that the approximation obtained in [12] and [14] was not the most accurate representation of the PHP and its PGFL. The authors in [15] then presented a more approximate solution with tight bounds, even though the solutions were still non-tractable. The PHP technique reported in [15] was adopted in the CRN in [16]. Interested readers are referred to these works for more details. From [15], $I_{sp} = \sum_{x_i \in \Psi_s \cap b^c(Y_o, D)} P_s h_{x_{sp}} ||x_{sp}||^{-\eta}$, where $b^c(Y_o, D)$ represents a disk of radius D centred on the PR Y_o. Conditioned on distance $||v||$, the $\mathcal{L}_{I_{sp}}$ can be expressed as:

$$\mathcal{L}_{I_{sp}}(s) \triangleq \mathcal{L}(\mu_s, P_s, s, ||v||, D)$$

$$= \exp\left(-\pi \mu_s \frac{(s P_s)^\tau}{sinc(\tau)}\right) \exp\left(\int_{||v||-D}^{||v||+D} \frac{2\pi \mu(r)}{1 + \frac{r^\eta}{s P_s}} r dr\right), \quad (9.23)$$

where $\mu(r) = \frac{\mu_s}{\pi} \cos^{-1}\left(\frac{r^2 + ||v||^2 - D^2}{2||v||r}\right)$.

9.4.2   Interference Modelling in the Secondary Network

Interference modelling in the secondary network is similar to the methods presented for the primary network. At any tagged SR y_o, the probability that the packet sent by its paired ST is successfully received is given as:

$$P_{suc}(y_o) = P(SINR_{y_o} > \theta_s), \quad (9.24)$$

where θ_s is the predefined threshold SINR for secondary transmissions. From Eq. (9.4) and (9.6), we know that:

$$P_{suc}(y_o) = P\left(\frac{P_s h_{x_{ss}} ||x_{ss}||^{-\eta}}{W + I_{ss} + I_{ps}} > \theta_s\right),$$

$$= P\left\{h_{x_{ss}} > \left(\frac{\theta_s ||x_{ss}||^\eta}{P_s} W + \frac{\theta_s ||x_{ss}||^\eta}{P_s} I_{ss} + \frac{\theta_s ||x_{ss}||^\eta}{P_s} I_{ps}\right)\right\}. \quad (9.25)$$

With the assumption of Rayleigh fading, the channel power gain $h_{x_{ss}} \sim \exp(1)$. Similarly, the Euclidean distance between the test secondary transmitter–receiver pair is also considered to be constant, that is, $r_s = ||x_{ss}||$. From this:

$$P_{suc}(y_o) = E\left\{\exp\left(\frac{-\theta_s ||x_{ss}||^\eta}{P_s} W - \frac{\theta_s ||x_{ss}||^\eta}{P_s} I_{ss} - \frac{\theta_s ||x_{ss}||^\eta}{P_s} I_{ps}\right)\right\}. \quad (9.26)$$

Taking $s = \frac{\theta_s ||x_{ss}||^\eta}{P_s}$, the expression in Eq. (9.26) is simplified as:

$$P_{suc}(y_o) = \exp(-sW)\mathcal{L}_{I_{ss}}(s)\mathcal{L}_{I_{ps}}(s). \quad (9.27)$$

Case 1 Consider a typical CRN where the PTs' distribution follows a homogeneous PPP Ψ_p of intensity μ_p and the STs' distribution follows an independent PPP Ψ_s of intensity μ_s. The derivation for I_{ss} can be obtained as interference from all active STs except the tagged ST and depends on a few assumptions in the secondary network. The LT of $I_{ss} = \sum_{x_i \in \Psi_s \backslash x_o} P_s h_{x_{ss}} ||x_{ss}||^{-\eta}$ can be upper-bounded at [12]:

$$\mathcal{L}_{I_{ss}}^{upper}(s) = \mathcal{L}(\mu_s, P_s, D). \tag{9.28}$$

The approximate expression for the $\mathcal{L}_{I_{ss}}$ can be obtained by considering the STs that are located outside the PUs' exclusion regions. This is given as:

$$\mathcal{L}_{I_{ss}}(s) = \mathcal{L}(\mu_s^a, P_s, s). \tag{9.29}$$

Finally, the expression for the LT of I_{ps} can also be obtained following:

$$I_{ps} = \sum_{X_i \in \Psi_p} P_p h_{X_{ps}} ||X_{ps}||^{-\eta},$$

and can be approximated as:

$$\mathcal{L}_{I_{ps}}(s) \triangleq \mathcal{L}(\mu_p, P_p, s, \bar{D}). \tag{9.30}$$

Note that these analyses are based on the receiver-centric scenario where any exclusion region of radius D is centred on each PR, indicating that any test ST within the network is thus located at a distance of at least D from the nearest PR. Since the distance between any primary transmitter–receiver is assumed to be r_p, while the distance between any secondary transmitter–receiver is r_s, then, it is clear to observe that the minimum distance between the nearest PT and the test ST is $D - r_p$. It becomes immediately clear that the nearest PT is at least a distance $\bar{D} = D - r_p - r_s$ from the test SR. It is also important to mention that when the analyses are obtained based on the transmitter-centric scenario where the exclusion regions are rather centred on each PT, $\bar{D}$ is slightly modified. In such a case, the location of the nearest PT is at least a distance $\bar{D} = D - r_s$ from the test SR.

Case 2 When the PTs are distributed following a homogeneous PPP Ψ_p^a of intensity $\mu_p^a = \mu_p \exp(-\mu_p \pi D^2)$ and the STs are distributed following a PHP Ψ_s^{php} of intensity μ_s^{php}, the exact form of the interference from the STs at a test SR is difficult to obtain because of the location-dependent thinning of Ψ_s to Ψ_s^{php}. As a result of this, only its approximate can be obtained. The approximate form of $\mathcal{L}_{I_{ss}}$ is given as:

$$\mathcal{L}_{I_{ss}}(s) = \mathcal{L}(\mu_s^{php}, P_s, s). \tag{9.31}$$

Similarly, the interference received at the test SR as a result of the activities of the PTs is dominated by interference from location $\bar{D}$, since the closest PT is at least a distance $\bar{D}$ from the tagged SR. The LT of I_{ps} under this scenario can be bounded at:

$$\mathcal{L}_{I_{ps}}(s) = \mathcal{L}(\mu_p, P_p, s, ||v||, \bar{D}). \tag{9.32}$$

There are other cases such as modelling the distributions of the PUs and the SUs following hardcore point processes. Such models ensure that the network considers both inter-network and intra-network dependence among the PUs and the SUs. The analyses for such models are presented in [16], while the details of PCP are discussed in [12].

The analyses presented thus far focussed on the underlay CRN model where the SUs are permitted to transmit within the same channels with the PUs. Hence, both intra-network and inter-network interference were considered in the modelling. In the overlay CRN model, the analyses can be simplified depending on the interest of the author. For instance, some authors considered inter-network interference to be negligible in the overlay CRN model. With such an assumption, the analyses become even more simplified.

9.5   Summary of the Chapter

The tool of SG is a very useful tool for interference management and control in any wireless network and its adoption for the CRN continues to receive serious attention. This is because the tool of SG has the ability to capture the locations of users in more practical terms, unlike the grid-based model. Despite the existence of various challenges when modelling interference using SG, new results are showing that tractable and accurate solutions can be obtained when channel parameters are carefully selected. It is believed that SG-based interference models can be significant towards effective and efficient resource allocation process in the CRN.

References

1. B.S. Awoyemi, B.T. Maharaj, A.S. Alfa, Resource allocation in heterogeneous buffered cognitive radio networks. Wirel. Commun. Mob. Comput. **2017**(7385627), 1–12 (2017)
2. J. Xu, J. Zhang, J. G. Andrews, On the accuracy of the Wyner model in cellular networks. IEEE Trans. Wirel. Commun. **10**(9), 3098–3109 (2011)
3. H.S. Dhillon, R.K. Ganti, F. Baccelli, J.G. Andrews, Modeling and analysis of K-Tier downlink heterogeneous cellular networks. IEEE J. Select. Areas Commun. **30**(3), 550–560 (2012)
4. C.H. Lee, C.Y. Shih, Y.S. Chen, Stochastic geometry based models for modeling cellular networks in urban areas. Wirel. Netw. **19**, 1063–1072 (2013)

5. T.V. Nguyen, F. Baccelli, A stochastic geometry model for cognitive radio networks. Comput. J. **55**(5), 534–552 (2012)
6. J.G. Andrews, F. Baccelli, R.K. Ganti, A tractable approach to coverage and rate in cellular networks. IEEE Trans. Commun. **59**(11), 3122–3134 (2011)
7. K.S. Gilhousen, I.M. Jacobs, R. Padovani, A.J. Viterbi, L.A. Weaver, C.E. Wheatley, On the capacity of a cellular CDMA system. IEEE Trans. Veh. Technol. **40**(2), 303–312 (1991)
8. H. ElSawy, E. Hossain, M. Haenggi, Stochastic geometry for modeling, analysis, and design of multi-tier and cognitive cellular wireless networks: a survey. IEEE Commun. Surv. Tutorials **15**(3), 996–1019 (2013)
9. D. Moltchanov, Distance distributions in random networks. Ad Hoc Netw. **10**(6), 1146–1166 (2012). http://www.sciencedirect.com/science/article/pii/S1570870512000224
10. P. Cardieri, Modeling interference in wireless ad hoc networks. IEEE Commun. Surv. Tutorials **12**(4), 551–572 (2010)
11. X. Song, C. Yin, D. Liu, R. Zhang, Spatial throughput characterization in cognitive radio networks with threshold-based opportunistic spectrum access. IEEE J. Select. Areas Commun. **32**(11), 2190–2204 (2014)
12. C. Lee, M. Haenggi, Interference and outage in Poisson cognitive networks. IEEE Trans. Wirel. Commun. **11**(4), 1392–1401 (2012)
13. U. Tefek, T.J. Lim, Interference management through exclusion zones in two-tier cognitive networks. IEEE Trans. Wirel. Commun. **15**(3), 2292–2302 (2016)
14. S. Kusaladharma, C. Tellambura, Secondary user interference characterization for spatially random underlay networks with massive MIMO and power control. IEEE Trans. Veh. Technol. **66**(9), 7897–7912 (2017)
15. Z. Yazdanshenasan, H.S. Dhillon, M. Afshang, P.H.J. Chong, Poisson hole process: theory and applications to wireless networks. IEEE Trans. Wirel. Commun. **15**(11), 7531–7546 (2016)
16. S.D. Okegbile, B.T. Maharaj, A.S. Alfa, Interference characterization in underlay cognitive networks with intra-network and inter-network dependence. IEEE Trans. Mob. Comput., 1–1 (2020)

Chapter 10
Deep Learning Opportunities for Resource Management in Cognitive Radio Networks

10.1 Introducing Machine and Deep Learning into Cognitive Radio Networks

Modern cognitive radio networks (CRN) is challenging traditional wireless networking paradigms by introducing learning and reasoning concepts, which are firmly stemmed into artificial intelligence (AI), in order to foster spectrum (resource) management [1]. This new space has allowed a plethora of potential applications such as cognitive wireless backbones and cognitive machine-to-machine devices to spatially and/or temporally reuse the wireless spectrum. Consequently, the CRN has been advocated as one of the most prominent technologies for contemporary spectrum management in modern wireless communications [2].

As already established, the CRN uses the notion of dynamic spectrum access (DSA), through opportunistic radio resource allocation (RA), to describe how to manage and optimise the utilisation of the limited spectrum resources in meeting the demands of unlicensed devices [3]. Overarching results in RA for the CRN have been published and have demonstrated higher sum-rate, as well as interesting trade-off in terms of delay-energy and throughput. However, with the acceleration towards beyond fifth-generation (5G) networks, where the internet-of-things (IoT) will be the primary deployment strategy, the delay-energy and throughput trade-off are no longer the only performance objectives to be achieved [4].

Furthermore, as mobile and wireless communication networks undergo a change in landscape, the signs are already discernible that spectrum scarcity will no longer be the only bottleneck to the advancement of wireless technology. With the emergence of compelling applications such as the IoT and its variants, such as the cognitive IoT (CIoT) and the cognitive internet of people, process, data and things (CIoPPD&T) [5], the spectrum scarcity problem will be paralleled with and increasing need for computational resources. These compelling applications will revolutionise wireless network operations by imposing many new challenges,

B. TJ Maharaj, B. S. Awoyemi, *Developments in Cognitive Radio Networks*, https://doi.org/10.1007/978-3-030-64653-0_10

including advanced channel modelling, low-latency requirement in large-scale hyper-dense connectivity, and much more.

As a result, the optimisation of wireless networks using AI strategies has become topical among researchers and more wireless networking solutions are being proposed to address these challenges. The success of *deep learning* (DL) in various fields, particularly in computational science, has recently stimulated increasing interest in applying it to address those challenges of modern communications. The DL, also known as deep neural learning, is a subset of *machine learning* (ML). Both DL and ML are AI functions with the goal of building systems that use intelligence to solve complex tasks [6].

Applications of DL in mobile and wireless networking come in two different forms; *the architectural design* and *the algorithmic design*. The architectural design breaks the classical model-based block design rule of wireless communications into one end-to-end communication system, while the algorithmic design manifests in a series of typical evolutionary techniques conceived for 5G networks and beyond. These evolutionary algorithms used in designing autonomous systems capable of solving complex problems are a goal that has been dreamed for decades and also the goal of the entire AI research community.

Evolutionary algorithms, more precisely deep architecture (that is, DL and deep reinforcement learning (DRL)) are bridging the digital and the physical divide by leveraging IoT and cyber-physical systems and striving towards ever more automation. However, when designing systems to make autonomous decisions in dynamic and distributed CRN, one requires quite exceptional computational resources to deal with the resulting volume of data that needs to be gathered and analysed [7]. This requires thorough and in-depth investigations into the essentials and intricacies of spectrum resource management and computational superiority. Thus, the combination of CRN, mobile edge computing, advanced predictive and prescriptive analytics, and AI have become of so much importance in achieving dynamic RA in modern wireless networks. This is certainly the most prominent reason for classifying the CRN among the top IoT evolutional technologies in recent times [8].

As modern technologies evolve, investigating the principles underlying the design and implementation of robust and pervasive networked computing systems has become the specific goal of most future mobile and wireless networks [9]. As a result of this new communication ecosystem landscape, current and future research on mobile and wireless communications has become an inter-disciplinary field. This field is shaped by different but interacting dimensions, and links the technological perspective closely to the social, economic and cognitive sciences.

Furthermore, what drives most current research activities in this resulting inter-disciplinary field is how to build reliable local algorithms based on local knowledge that can derive globally-emergent system characteristics such as reliability, availability, efficient resource utilisation, quality assurance, etc. The overall objective of this chapter is to discuss how the application of AI techniques through model-based DL can help in constructing predictive models that are able to explain the behaviour of future resource characteristics using the underlying physical process of mobile and wireless network dynamics, particularly for CRN applications.

In all, the techniques discussed in this chapter can be thought of as a hierarchy of concepts, that is, using simple concepts to build more complex concepts by structurally putting simple concepts in a hierarchical form, making up what is referred to as the *deep architecture*. Deep architecture are systems composed of multiple levels of non-linear operations, such as neural networks (NN), with many hidden layers [10]. The deep structure of concepts arranged in a hierarchical form can be thought of as the reason for referring to the field as DL.

10.2   Understanding Machine Learning and Deep Learning

Several ML techniques have progressed outstandingly since the proposition of the Turing Machine by Alan Turing. To date, ML is the most common AI technique used for processing Big Data using self-adaptive algorithms that get better analysis and patterns with experience or with new added data [11]. While ML algorithms build their analysis with data in a linear way, DL has become more influential because of its use of a hierarchical level of artificial neural networks (ANNs) to carry out its processes. The hierarchical function of DL systems enables machines to process input data comprised of features having a non-linear approach to deliver results in the form of predictions, as illustrated in Fig. 10.1.

Each ANN in Fig. 10.1 consists of layers of simple computational nodes that work together to munch through data and deliver an output. The input data, which encompass a lot of things such as numbers, words, images, etc. are fed into the ML algorithms which use statistics to compute their results. The first layer of the ANN

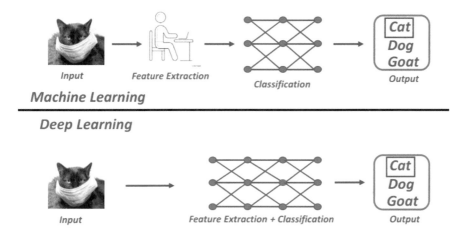

Fig. 10.1 Comparison between deep learning and machine learning

processes a raw data input and passes it on to the next layer as output. The next layer processes the previous layer's information by including additional information such as the weights and biases. This process continues across all levels of the ANN until the best output is determined. The main difference between ML and DL is that *ML separates feature selection and extraction from classification, while DL performs both feature extraction and classification in a single neural network (NN) using end-to-end learning.*

10.3   Incorporating Deep Learning in Wireless Networks

Traditional wireless transmission relies on accurate mathematical models of the wireless channels to design channel estimation algorithms or channel feedback schemes. However, DL usually does not rely on such mathematical models for its tasks, and is particularly beneficial in model-deficit and algorithmic-deficit problems. Model and algorithmic deficiency entail the absence of and/or insufficiency in accurate and well-established mathematical models and algorithms needed to help solve a particular problem [12].

In DL, logical formulas written in temporal logic can be used for training the weights of a feed-forward-feedback network using the structure shown at the bottom of Fig. 10.1. A linear transformation projects the input data into a *space* (which acts as an intermediate layer) where it becomes linearly separable. This intermediate layer is referred to as a hidden layer and the premise of DL is that there are substantial benefits to using many hidden layers. The number of hidden layers is influenced by how complicated the input distribution is. The more complicated the input distribution, the more the number of hidden layers required. Thus, a hierarchical function of a feed-forward-feedback NN depends on the size of the problem to be solved. This also influences the computational capacity the network will require to model it.

As traditional wireless transmission tends to design each module of the communication system separately using mathematical derivations, DL usually train all the parameters of the deep neural network (DNN) as a whole. For example, in order to have perfect channel knowledge to provide enough insights for the understanding of the system, channel estimation has to be included in the expressions of the training algorithm. In this way, the achievable rate can be obtained by accounting for the bandwidth and the time- and frequency-spacing used by the network.

10.4   Training a Deep Learning Model

The DL models use NNs to learn a mapping function from inputs to outputs and accuracy is achieved by updating the weights of the NN in response to the errors that the model makes during training. Updates are then made by adjusting the weights in order to continually reduce the errors until an acceptable model is achieved. This process is the most challenging part of using DL algorithms and is also the most time consuming. An NN can be defined using the function $f : R^N \rightarrow R^L$, where N is the size of input vector x and L is the size of the output vector $f(x)$ such that, in matrix notation, $f(x)$ is defined as [13];

$$f(x) = \Phi(W(x) + b),\tag{10.1}$$

where $\Phi(x) = \frac{1}{1+e^{-x}}$ is the activation function, W denotes the weight matrix, b represents the bias variable without which the given layer will not produce an output that differs from zero in the next layer. The activation function can be viewed as a logistic regression classifier where the input is first transformed using a learnt non-linear transformation Φ, as shown in Eq. (10.1) [14]. Typical choices of activation functions include the hyperbolic tangent $f(x) = \tanh(x) = \frac{e^x - e^{-x}}{e^x + e^{-x}}$, the rectified linear unit (ReLU) $f(x) = f(x)$, or the Boltzmann softmax $f(x) = \frac{e^{-x}}{\sum_{j=1}^{K} e^{-x}}$.
For most applications in wireless communications, the parametric ReLU, that is, $\max(\cdot, 0)$ is used as the hidden layer activation function to learn the appropriate value of the first argument, as shown in Fig. 10.2.

In the example application (Fig. 10.2), the activation function of the output layer is usually the Boltzmann Softmax in most wireless communications applications. The Softmax function is typically used to classify outputs into multiple categories, such as different transmission powers arranged in ascending order. In order to train

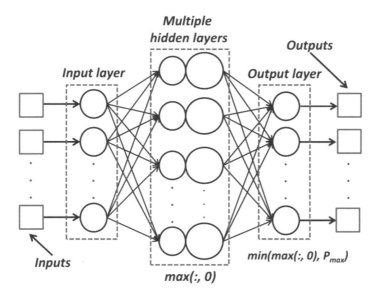

Fig. 10.2 A normal feed-forward multi-layer perceptron

a multi-layer perceptron (MLP), the set of parameters to learn is the set $\mathbf{w} = \{W, b\}$, while the forward propagation is defined as follows:

$$
\begin{aligned}
h_1 &= \Phi(w_1 * x); \\
h_2 &= \Phi(w_2 * h_1); \\
h_3 &= \Phi(w_3 * h_2); \\
y &= \Phi(w_4 * h_3),
\end{aligned}
\tag{10.2}
$$

where h_1 denotes the index of the output of the first hidden layer, w_1 represents the weight index of the weights of the first hidden layer, $(\cdot * \cdot)$ denotes a convolution operation, and y is the output. Obtaining the gradients $\partial \mathcal{L}/\partial \mathbf{w}$ can be achieved through a backward propagation algorithm, which is a special case of the chain-rule of derivation given as follows:

$$
\mathcal{L}(w) = w_1 - \alpha \frac{\partial L(w)}{\partial w_1} w_2 - \alpha \frac{\partial L(w)}{\partial w_2} w_3 - \alpha \frac{\partial L(w)}{\partial w_3} w_4 - \alpha \frac{\partial L(w)}{\partial w_4},
\tag{10.3}
$$

where the term α is the step size or the learning rate. The loss function $\mathcal{L}(w)$ is the measure of the difference between the output $\mathbf{y}$ and the actual ground $\mathbf{y}^*$. Therefore, the training objective is based on obtaining the best weights $\mathbf{w}$ that minimises the $\mathcal{L}$ function. The parameters of the model are learned using stochastic gradient descent (SGD) with mini-batches, such that the gradient of the $\mathcal{L}$ function is computed over the weight of the last hidden layer, and the weights are updated as follows:

$$
w_4 = w_4 - \alpha \frac{\partial \mathcal{L}(w)}{\partial w_4},
\tag{10.4}
$$

which repeats until the gradient descent eventually results to a set $\mathbf{w}$ that minimises $\mathcal{L}$. Using this computational technique, DL has provided remarkable capabilities and advancements in many areas such as pattern recognition, image processing, natural language processing, etc.

10.5   Application of Deep Learning in Spectrum Management

The application of DL strategies in wireless communications require features with specific characteristics in order for them to function properly. In as much as the DL strategies are predominantly data driven, in wireless networking problems, the data-driven approaches do not replace but rather complement traditional design techniques. Thus, the need for signal processing techniques, such as mathematical programming and nature-inspired techniques, whose objective is to prepare proper input dataset, compatible with ML algorithm requirements, will always be there.

Even though there is an explosive proliferation of DL applications in wireless communications, by contrast, the application of DL strategies in spectrum management is not straight-forward, except for spectrum sensing. Although DL strategies enable for the creation of machines that have high accuracy in specific tasks, they are still limited in making certain decisions. They are relatively weak in problems beyond classification and dimensionality reduction, and this has limited their applicability in the wireless network economics involved in spectrum access.

Spectrum sensing algorithms for the orthogonal frequency division multiplexing (OFDM) signal based on DL and covariance matrix graphs were presented in [15], where the outstanding capability of DL in image processing was exploited and OFDM signals were analysed using the structural characteristics of the covariance matrix. A stacked auto-encoder (SAE) was also applied on time domain signals in spectrum sensing, with the input dataset consisting of time domain signals. With conventional OFDM suffering from issues of noise uncertainty, time delay and carrier frequency offsets, the framework developed was to help address those specific challenges of the OFDM-based networks.

Further, a blind spectrum sensing method based on DL was studied in [16] to help improve spectrum sensing in low signal-to-noise ratio (SNR) situations where prior information of the licensed user was not available. In that application, three kinds of NNs were used together; convolutional NNs, long short-term memory (LSTM) NNs and fully connected NNs. This resulted in improved performance compared to the traditional energy detector, especially in low SNR regimes. The effect of different LSTM memory layers was also analysed and an exploration of the improved performance by the DL-based detector was carried out. The motivation for using several LSTM layers was to establish a more efficient model of the probability distribution of the observed sequence of hidden Markov models, extract the timing features of the signals, and to distinguish the signal and noise from the timing regularity of the input data.

A spectrum occupancy reconstruction technique for missing spectrum data imputation in collaborative spectrum sensing was studied in [17], where the objective was to reconstruct an incomplete spectrum sensing data matrix. Represented as a plenary grid on a Markov random field (MRF), the problem was formulated as a magnetic excitation state recovery problem, and the SGD method was applied to solve the matrix factorisation. The learned statistics interfaced onto sparse approximation on the physical layer using techniques such as the mean squared error (MSE) provided proof that the sparse approximation performs better than singular value decomposition techniques. The above-mentioned works are some of the recent works that have introduced DL into spectrum management in the CRN.

10.6 Deep Reinforcement Learning

Since spectrum access is a goal-directed phenomenon and DL algorithms cannot solve them alone, deep reinforcement learning (DRL) techniques are usually exploited to extract rich features from the CRN environment, and to enable/achieve the required level of intelligence. The key idea of DRL is to use deep neural networks (DNNs) to represent Q-networks, and to train this Q-network to predict future reward.

10.6.1 Training the Deep Reinforcement Learning Model

A DRL model is trained using a Q-learning algorithm whose task is to interact with both the environment and the model. For this reason, the algorithm that runs in a DRL model is known as the DRL agent [18]. The agent may arrive at different environmental scenarios known as *states* $s \in S$ by performing *actions* $a \in \mathcal{A}$, where S and $\mathcal{A}$ are the state and action spaces, respectively. Actions lead to *rewards* $r \in \mathcal{R}$ which could be positive or negative. Each state s_t within the environment is a consequence of a previous action a_t, which in turn results in the next state s_{t+1}.

Storing all the information, however, becomes infeasible and this is resolved by assuming that the sequence of states follow a Markov property. This means that each state depends solely on the previous state and the transition from that state to the current state. This intuition is the one behind Markov decision processes (MDPs) and for the purpose of decision-making in different transmission modes, channel transition probabilities are described using MDPs. If the expected reward at each action at time step t is known, this would essentially be like a cheat sheet for the agent, since the agent would know exactly which action to take to eventually obtain the maximum reward [19]. This total reward is also called the Q-value and is formalised as follows:

$$Q(s, a) = r(s, a) + \beta \max_{a \in \mathcal{A}} Q(s_{t+1}, a), \qquad (10.5)$$

where the term $0 < \beta < 1$ is the discount factor that controls the contribution of future rewards, such that adjusting the value of β either diminishes or increases contribution of future rewards. An example of a DRL formulation using a DNN is shown in Fig. 10.3.

In Fig. 10.3, the elements of the state set are the channel gains g_t and the current reward R_t. A top-level Q-value function learns a policy over the defined goals and prescribes actions to satisfy the given goals. This structure allows for a flexible specification of goals and provides an efficient space for exploration in complicated environments such as the CRN.

The Q-value obtained from being at the state s and performing an action a is the immediate reward $r(s, a)$ plus the highest Q-value possible from the next state s_{t+1}. Since Eq. (10.5) is a recursive equation, assumptions on all Q-values are made in the beginning and with experience, the equation converges to the optimal policy. The update equation is implemented as follows:

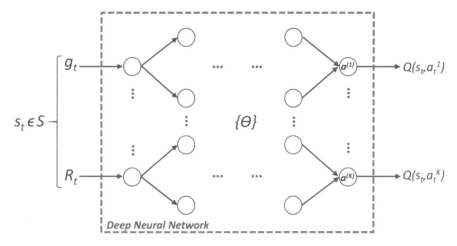

Fig. 10.3 Illustration of service classification using a DNN

$$Q(s_t, a_t) \leftarrow Q(s_t, a_t) + \alpha \left[R_{t+1} + \beta \max_{a \in \mathcal{A}} Q(s_{t+1}, a) - Q(s_t, a_t) \right], \qquad (10.6)$$

where $0 < \alpha < 1$ is the learning rate, which determines to what extent newly acquired information overrides the old one. The estimated Q-function for a single state-action pair is subsequently updated by following a gradient descent step to minimise the loss. This results in the update as follows:

$$Q_{t+1}(s, a) = Q_t(s, a) - \alpha \frac{\partial \mathcal{L}(s, a)}{\partial Q_t(s, a)}, \qquad (10.7)$$

where $\alpha > 0$ is the learning rate. After gradient evaluation and emerging terms, Eq. (10.7) becomes

$$Q_{t+1}(s, a) = (1 - \alpha) Q_t(s, a) + \alpha \left[\mathbf{1}^T \mathbf{c}(s, a) + \beta \min_{a \in \mathcal{A}} Q_t(s, a) \right], \qquad (10.8)$$

where $\beta > 0$ is the discount factor. Equation (10.8) defines a minimisation of the cost function $c(\cdot)$, as exemplified by the $\min_{a \in \mathcal{A}}$ in the last term. However, an equivalent maximisation problem is defined as follows:

$$Q_{t+1}(s, a) = (1 - \alpha) Q_t(s, a) + \alpha \left[\mathbf{1}^T \mathbf{r}(s, a) + \beta \max_{a \in \mathcal{A}} Q_t(s, a) \right], \qquad (10.9)$$

where the a is the object of choice in the strategy interaction process, such that a SU that has chosen a receives an instantaneous pay-off $G(a, v) = \mathbf{c}(s, a) = \mathbf{r}(s, a)$, where v is the choice action for a neighbouring SU, and G is the game matrix.

10.6.2 Application of Deep Reinforcement Learning in Spectrum Management

In spectrum management, DRL offers a multitude of approaches from strategic interaction processes to the multi-armed bandit (MAB) schemes. Strategic inter-action processes are game theoretic techniques, commonly used in economics, for modelling interactions between two or more players in situations involving a set of rules and outcomes [20]. The MAB scheme is a huge problem space with many dimensions along which the models can be made more expressive and closer to reality. The MAB scheme is a multi-agent reinforcement learning scheme from probability theory that falls into the broad category of stochastic scheduling [21].

In the MAB scheme, a fixed limited set of resources must be allocated between competing (alternative) choices in a way that maximises their expected gain. The objective of the MAB problem is to search for an allocation of channels for all users that maximises the expected sum throughput. The problem is usually formulated as a combinatorial MAB, in which each arm corresponds to a matching of the users to channels. However, if the properties related to each choice are only partially known at the time of allocation, and may become better understood as time passes, the transition probabilities are described using partially observable MDPs (POMDPs).

The strategy interaction process combines graph theory and game theory into the reinforcement learning formulation, similar to the one proposed in [18], where a DRL strategy was used to maximise the system utility in terms of improving system throughput and optimising energy efficiency in cognitive relay networks. An apprenticeship DRL scheme for energy-efficient cross-layer routing was studied in [22], where dynamic adjustment rating was employed to compress the huge action space to guarantee energy efficiency. Dynamic adjustment rating was used to efficiently regulate the transmission power using a multi-level transition mechanism. This technique confirmed that dynamic adjustment rating achieves higher energy efficiency, reduced latencies and achieved better packet delivery ratios.

An online learning policy for distributed CRN that takes into account the channel availability criteria, together with the quality metric related to inter-cell interference, was designed as a multi-user Markov MAB problem in [23]. In the work, the SUs selfishly collect a priori unknown rewards by selecting a channel without any information exchange between them. Using a relentless QoS upper-bound confidence, the SUs were able to select the best channel to transmit. In [24], a blind spectrum selection problem for SUs with poor sensing abilities was studied based on a MAB framework for medium access in decentralised CRN. Since the channel statistics were unknown a priori, taking hand-off delays as a cost, the problem was formulated using POMDPs.

A stochastic multiplayer MAB problem was studied in [25] where several players pull arms simultaneously and collisions occur if one of them is pulled by several players at the same stage. A decentralised algorithm that contradicts the existing lower bound for that problem was found to achieve the same performance as the centralised one. This was made possible by hacking the standard model and constructing a communication protocol between players that deliberately enforces collisions, allowing them to share their information at a negligible cost. This motivated the introduction of a more appropriate dynamic setting without sensing, where similar communication protocols were no longer possible.

Despite the great promises of MAB schemes, the striking drawback of these schemes is that the number of arms grow super-exponentially as the permutation between channels and SUs. In this case, a matching-learning algorithm with polynomial storage and polynomial computation per decision period for this problem is required. To circumvent this issue, DRL algorithms that develop a real-time adaptive policy for computational RA of multiple users was proposed by Google DeepMind [26]. This was a pioneering contribution in DRL which successfully combined convolutional neural networks (CNNs) and Q-learning to train reinforcement learning agents with just a few inputs. The newly-formed deep Q-networks (DQNs) are hierarchical structures that are capable of storing policies that previously resulted in better rewards, and are thus perfect for networks that require long-term planning and decision-making processes such as the CRN.

10.7 Hierarchical Deep Architectures for Cognitive Radio Networks Applications

Hierarchical structures combine DL and reinforcement learning into a hierarchy of functions with the objective of attaining the knowledge to be used in learning hierarchical decomposition of spatial environments. In the context of deep architecture, DQNs and double deep Q-networks (DDQNs) are a step ahead of all AI strategies in terms of achieving better decision-making. Their capability to store information for future replay makes them suitable for long-term planning. Their architecture is known as the actor-critic architecture since the learning is always on-policy, meaning that the critic must learn about and critique whatever policy is followed by the actor [27]. In their policy structure, the actor is used to select actions, while the critic, which criticises the actions made by the actor, gives the estimated value function.

10.7.1 General Model of a Hierarchical Deep Architecture

Since the actor-critic terminology is a high-level concept, a more specific terminology borrowed from control systems is being used instead. With this, the actor is referred to as the controller while the critic is called the meta-controller. A low-level abstraction of a hierarchical deep architecture that uses a two-stage hierarchy consisting of two temporal abstractions, the controller and meta-controller is illustrated in Fig. 10.4.

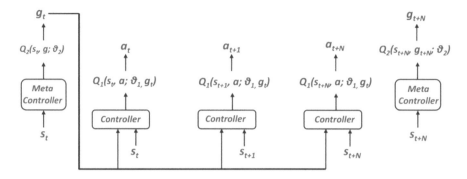

Fig. 10.4 A hierarchical deep architecture consisting of a two-stage hierarchy, the controller and the meta-controller

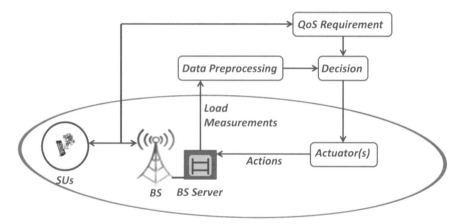

Fig. 10.5 A high-level abstraction of a hierarchical deep model employing a model-based reinforcement learning formulation

Separate instructions are used in both the controller and the meta-controller. The meta-controller looks at the raw states $s_t \in S$, where S represents the set of all possible states. It then produces a policy θ over goals by estimating the action-value function $Q_2(s_t, g_t; \theta_2)$ to maximise the future rewards. The meta-controller executes plans to sequence the operations of the system, which corresponds to asynchronous sensor data interrupts in real time, and issues actuator commands to a real-time controller module. In intelligent control, the meta-controller receives the state s_t and chooses a goal $g_t \in G$, where G denotes the set of possible current goals. Furthermore, the software structure of the actuator translates the set of possible current goals into actions, as seen later in Fig. 10.5.

The controller takes in the current state s_t and the current goal g_t. It then selects an action $a_t \in \mathcal{A}$, where $\mathcal{A}$ is the set of all possible actions. Further, it produces a policy over actions by estimating the action-value function $Q_1(s_t, a_t; \theta_1, g_t)$ to solve the predicted goal by maximising the expected future reward. If the goal is reached, the controller provides a positive reward and decides when to end an episode. This means that the goal can remain in place for the next few time steps t, either until it is achieved or a terminal state is reached. It does this by taking in states s_t. The decision module is responsible for evaluating whether a goal has been reached or not, and provides an appropriate reward $r_t(g)$ to the controller. The agent receives sensory observations and produces actions a_t. The meta-controller then chooses a new goal g and the entire process is repeated [28].

10.7.2 Application of Hierarchical Deep Reinforcement Learning in Cognitive Radio Networks and Edge Computing

With the advent of edge computing, the integration of the CRN with edge computing comes with the requirement to simultaneously handle spectrum management and processing requirements. This integration has also been met with the relentless push to make network operations more intelligent in order to fully unleash the potentials of wireless Big Data, which entailed pushing the frontiers of AI to the network edge. In order to present the CRN without any loss of generality, a CRN where SUs communicate via a base station (BS) over a time-varying fading channel in the presence of PUs is assumed.

The DRL agent, which resides within each device, subsumes the resource management for the entire CRN and is assumed to have knowledge of the coding scheme employed. Given a set of signals, we seek a framework that leads to the best representation for the transition function from the current state to the reward to the next state. In a time-varying channel context for CRN under strict sparsity constraints, the prerequisite for learning the wireless channel is its sparsity representation. The combination of data analytics with ML techniques promises to be a possible pathway towards achieving both QoS and energy efficiency objectives, and is an essential step towards managing the high level of heterogeneity that comes with beyond 5G networks. By obtaining tentative operating points for network equipment, drive towards learning and operating over different levels of temporal abstraction is a key to solving some of the CRN challenges. A detailed DL modelling approach which characterises the underlying process as a hierarchical architecture is illustrated in Fig. 10.5.

Figure 10.5 shows a high-level orchestration of a hierarchical deep architecture that integrates hierarchical functions of data analytics, reinforcement learning and model-based reinforcement learning. This model-based, data-driven, transfer learning-based approach follows the same approach described in Fig. 10.4. Assuming that the Poisson point process is sufficiently accurate to account for the distribution of cellular BSs, the focus is then shifted from the impact of the spatial distribution of cellular BSs to the power consumption model of a single BS. In this case, the whole RA problem can be divided into two stages; the opportunistic spectrum access which can be solved using traditional optimisation methods, and opportunistic computing which can be solved using DL techniques.

In the opportunistic computing stage, the BS processor can be viewed as a hybrid switching system whose task is to search and find its optimal operating state using control actions that are derived to drive a DL model. Here, a single edge processing node (BS) is shown where opportunistic spectrum access and opportunistic computing take place. The transitions between the BS operating modes are either triggered by events or by the passage of time. For example, if the transmission queue is empty, the processor is idle and saves power. However, when events arrive, the processor is switched to active mode with little time overhead. If the processor stays idle beyond some threshold duration, it is placed in the sleep mode for a specified period.

The hierarchical structure so far described is able to handle complex discrete stochastic decision processes with stochastic transitions in spectrum management for the CRN. In [13] the DL model of choice was a stacked auto-encoder (SAE). However, the choice of the DL model to use is application-dependent. The works that have attempted to integrate edge computing with the CRN, such as in [29], are still very few and far between, making it a very promising open research area.

10.7.3   Application of Hierarchical Deep Reinforcement Learning in Cognitive Radio Networks Energy Management

The combination of signal processing techniques with Big Data streams is one of the key features in 5G mobile networks. Furthermore, combining deep architecture, signal processing with Big Data increases the intelligence and reliability of decision-making and, consequently, unleashes the full potential of beyond 5G technologies such as the CRN and the IoT. This is similar to human societies, where there is a collective intelligence that belongs to everybody (in this case, device intelligence), and individual intelligence that belongs to the cloud (in this case, cloud intelligence). In order to implement this kind of intelligence, we need to state that each network device has a specific intelligence. In this way, every network device can access the cloud intelligence by connecting to it.

Constrained energy management solutions are made possible by making deep architecture compatible for future wireless communication networks which distribute the intelligence throughout the whole network. Since the intelligence does not reside in one place, but instead, it is distributed across network devices, it is interesting to observe that this approach resembles the way in which human knowledge is developed. The distributed RA technique is illustrated in Fig. 10.6.

Figure 10.6 highlights the use of a deep architecture that combines opportunistic spectrum access with opportunistic computation. Here, the allocation of computa-

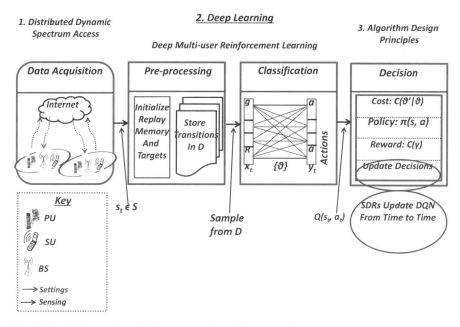

Fig. 10.6 An illustration of the implementation of deep learning in multi-user spectrum management for CRN applications

tional resources uses a DRL agent in mobile edge-computing networks that operates to support low-latency and energy-efficient communications. Using this technique, energy-efficient operation of cognitive radio devices and network infrastructure can be addressed together with the delay violation probability. In such an application, the predictor has to be designed based on traffic models that are as realistic as possible. However, since network traffic prediction is a complex non-linear system and it cannot fully reflect the variety of regulations by using a single model with higher prediction precision, the user behaviour is used to describe artificial traffic generators using chaotic maps.

Chaotic maps employ Poisson distribution and exponential distribution in the CRN. The artificial traffic generator feeds into a recurrent generalised Q-learning algorithm, which inherits the data and predicts the future behaviour of CRN traffic. As can be seen in Fig. 10.6, there are three modules required for this architecture.

1. **Distributed Dynamic Spectrum Access:** Data acquisition has recently become a critical issue and is currently a major bottleneck in RA for the CRN. As the IoT continues to steer operations and ML becomes widely used in wireless networking, designers are constantly confronted with the need for enough data to drive IoT applications. The radio environment in the CRN context consists of the wireless channel, the transmission buffer capacity, and the radio signal level at various points in the physical space as a function of time and frequency. The cognitive radio device collects this information and uses it as the state $s \in S$ of the CRN. The strength of accurate characterisation of the underlying process,

followed by the application of ML processes such as data mining, is the foundation of the overall decision-making objective. A data-driven proxy modelling approach and statistical learning for obtaining spectrum occupancy, like the one reported in [17], is a good starting point for the data acquisition process. In line with the developments in Big Data processing techniques, technologies to collect them from numerous distributed sources have also enhanced significantly. From distributed dynamic spectrum access systems such as sensor networks featured by software defined radio standard software, recurrent generalised Q-learning are used to extract useful information for analysing network behaviours that are relevant for spectrum management purposes.

2. **Deep Multi-User Reinforcement Learning:** The agent is defined as the algorithm running within the infrastructure, that is, a DRL algorithm running in the user device or the BS is called a DRL agent. The agent explores new behaviour that could help it to solve tasks posed by the environment. This process is problem-specific, but in order to satisfy curiosity, only the processes of data pre-processing and classification are discussed.

   • **Data Pre-processing:** Data pre-processing is a data mining technique that transforms raw data into an understandable or usable format. All data need to be pre-processed before it can be sent to a model. The data pro-processing in this context is done through importing libraries using a technique called *experience replay*. Experience replay is a DQN technique where historical data stored in memory buffers of distributed systems for future replay is sampled for training the system [30]. However, the sampling of the data from the memory D for correct replay is the one that poses a number of specific resource management related challenges. This is because the reliability of data streams plays a key role in how much history can be stored in the memory buffers. It is for this reason that the pre-processing and classification are combined to form the DL block to perform deep multi-user reinforcement learning in Fig. 10.6.

   • **Classification:** Classification is a supervised learning approach in which the algorithm learns from the input data and then uses the learned information to classify new observations. Through service class classification and analysis, services are classified into subsets of similar resources and performance requirements. Common classifiers that are also used in DL include *logistic regression*, *naive Bayes*, *nearest neighbour* and *support vector machines*, to name a few. In this application, and based on the pre-processed data, the objective of the classification task is to divide tasks into classes with similar resource requirements and performance characteristics in order to allocate available resources efficiently. The choice of classifier affects the goal of RA strategy, but there are versatile techniques such as the auto-encoder (AE) that can perform well for this application. In this case, the input to the AE can be a replay that has the decision based on the QoS requirements and the achievable QoS. The output of this task contains the actions $a_t \in \mathcal{A}$ that need to be executed and computed via a policy network $\{\theta\}$, as seen in Fig. 10.3.

3. **Algorithm Design Principles:** The design principles involve the final decision that needs to be executed by the algorithm, and this is usually the RA step that is a little difficult to achieve in a dynamic environment. At each time step, the DQN follows algorithm design principles that are defined by a cost function $J(\theta'|\theta)$, policy $\pi(s, a)$, and reward $R(\gamma) = \log_2(1 + \gamma)$, which the software defined radio will update after every episode (γ represents the SINR). At each instant, the software defined radio uses Open-Flow to access information about the queue lengths, and depending on the policy of its administrative domain, this information is passed onto the server to launch computational equipment. Based on the queue lengths and the incoming traffic load, $R(\gamma)$ is maximised while the cost $J(\theta'|\theta)$ is minimised. Then, the global solution created from this state of affairs can be stored in memory for future replay. In terms of the combination of opportunistic access and opportunistic computation, the decision-making will have to consider the cost, policy, and rewards which will be updated by the software defined radio from time to time.

So far, this technique has been applied to robotic control [31]. More sophisticated algorithms that utilise DQNs for constrained energy management in CRN can be developed using deep hierarchical DRL with prioritised experience replay. Future communication systems, which will be software defined radio-based and controlled by ML algorithms, can potentially benefit by managing communications system resources by monitoring performance functions with common dependent variables that result in conflicting goals using this technique. Since the uncertainty in the performance of thousands of different possible combinations of radio parameters makes the trade-off between exploration and exploitation in reinforcement learning much more challenging, this technique should make it possible for the system to spend as little time as possible on exploring actions, and whenever it explores an action, it should perform at acceptable levels, most of the time.

10.8   Summary of the Chapter

This chapter has explored and investigated a few application examples of deep architecture in solving RA problems in the CRN. The DL strategies discussed have found applications and achieved great results in the field of spectrum sensing, where the objective is a separation of one hypothesis from another. However, due to the involvement of different contextual hierarchies, DL algorithms still have limitations in achieving the best improvements in problems involving spectrum access, where the environmental states, consisting of the wireless channel, users, and the transmission buffer queues in both the user and the BS, need to be considered simultaneously. As discussed in the chapter, some apparent shortcomings in DL approaches are being addressed by the DRL approaches, with several key areas of application still requiring further investigations.

References

1. E. Meshkova, J. Riihijarvi, A. Achtzehn, P. Mahonen, Exploring simulated annealing and graphical models for optimization in cognitive wireless networks, in *Proceedings of the IEEE GLOBECOM* (2009), pp. 1–8
2. B. Awoyemi, B. Maharaj, A. Alfa, Optimal resource allocation solutions for heterogeneous cognitive radio networks. Digital Commun. Netw. **3**(2), 129–139 (2017). http://www.sciencedirect.com/science/article/pii/S2352864816301043
3. G.I. Tsiropoulos, O.A. Dobre, M.H. Ahmed, K.E. Baddour, Radio resource allocation techniques for efficient spectrum access in cognitive radio networks. IEEE Commun. Surv. Tutorials **18**(1), 824–847 (2016)
4. G. Soós, D. Ficzere, P. Varga, User group behavioural pattern in a cellular mobile network for 5G use-cases, in *NOMS 2020 - 2020 IEEE/IFIP Network Operations and Management Symposium* (2020), pp. 1–7
5. L. Dai, R. Jiao, F. Adachi, H.V. Poor, L. Hanzo, Deep learning for wireless communications: an emerging interdisciplinary paradigm. IEEE Commun. Mag., 05 (2020)
6. T. O'Shea, J. Hoydis, An introduction to deep learning for the physical layer. IEEE Trans. Cognitive Commun. Netw. **3**(4), pp. 563–575 (2017)
7. i Scoop, 5G and IoT: the mobile broadband future of IoT (2020). https://www.i-scoop.eu/internet-of-things-guide/5g-iot/
8. Ericsson, Discussing challenges in the internet of security (2020). https://www.ericsson.com/en/blog/2020/4/five-leaders-discuss-securing-iot
9. B. Blanco-Filgueira, D. García-Lesta, M. Fernández-Sanjurjo, V.M. Brea, P. López, Deep learning-based multiple object visual tracking on embedded system for IoT and mobile edge computing applications. IEEE Int. Things J. **6**(3), 5423–5431 (2019)
10. Y. Bengio, Learning deep architectures for AI. Found. Trends Mach. Learn. **2**(01), 1–55 (2009)
11. Y. Roh, G. Heo, S. Whang, A survey on data collection for machine learning: a Big Data - AI integration perspective. IEEE Trans. Knowl. Data Eng. **PP**(10) (2019)
12. M.C. Hlophe, A model-based deep learning approach to spectrum management in distributed cognitive radio networks. Ph.D. Dissertation, University of Pretoria (2020)
13. M.C. Hlophe, B.T. Maharaj, Qos provisioning and energy saving scheme for distributed cognitive radio networks using deep learning. J. Commun. Netw. **22**(3), 185–204 (2020)
14. R. Greve, E. Jacobsen, S. Risi, Evolving neural turing machines for reward-based learning, in *Proceedings of the Genetic and Evolutionary Computation Conference* (2016), pp. 117–124
15. M. Zhang, L. Wang, Y. Feng, H. Yin, A spectrum sensing algorithm for OFDM signal based on deep learning and covariance matrix graph. IEICE Trans. Commun. **E101.B**(05), 2435–2444 (2018)
16. K. Yang, Z. Huang, X. Wang, X. Li, A blind spectrum sensing method based on deep learning. Sensors **19**(10), 2270. https://doi.org/10.3390/s19102270
17. M.C. Hlophe, S.B.T. Maharaj, Spectrum occupancy reconstruction in distributed cognitive radio networks using deep learning. IEEE Access **7**, 14294–14307 (2019)
18. S. Liu, K. Hu, W. Ni, Z. Xu, F. Wang, Z. Wan, A cognitive relay network throughput optimization algorithm based on deep reinforcement learning. Wirel. Commun. Mob. Comput. **2019**(2731485), 1–8 (2019)
19. H. Li, T. Wei, A. Ren, Q. Zhu, Y. Wang, Deep reinforcement learning: framework, applications, and embedded implementations, in *Proceedings of the IEEE/ACM International Conference on Computer-Aided Design (ICCAD)* (2017), pp. 847–854
20. W. Zhou, J. Li, Y. Chen, L. Shen, Strategic interaction multi-agent deep reinforcement learning. IEEE Access **8**, 119000–119009 (2020)
21. Y. Gai, B. Krishnamachari, R. Jain, Learning multiuser channel allocations in cognitive radio networks: a combinatorial multi-armed bandit formulation, in *2010 IEEE Symposium on New Frontiers in Dynamic Spectrum (DySPAN)* (2010), pp. 1–9

22. Y. Du, Y. Xu, L. Xue, L. Wang, F. Zhang, An energy-efficient cross-layer routing protocol for cognitive radio networks using apprenticeship deep reinforcement learning. Energies **12**(07), 2829 (2019)
23. N. Modi, P. Mary, C. Moy, Qos driven channel selection algorithm for cognitive radio network: Multi-user multi-armed bandit approach. IEEE Trans. Cognitive Commun. Netw. **3**(1), 49–66 (2017)
24. Y. Chen, H. Zhou, R. Kong, L. Zhu, H. Mao, Decentralized blind spectrum selection in cognitive radio networks considering handoff cost. Fut. Int. **9**(03), 10 (2017)
25. E. Boursier, V. Perchet, SIC-MMAB: synchronisation involves communication in multiplayer multi-armed bandits. CoRR, vol. abs/1809.08151 (2018). http://arxiv.org/abs/1809.08151
26. V. Mnih, K. Kavukcuoglu, D. Silver, A. Rusu, J. Veness, M. Bellemare, A. Graves, M. Riedmiller, A. Fidjeland, G. Ostrovski, S. Petersen, C. Beattie, A. Sadik, I. Antonoglou, H. King, D. Kumaran, D. Wierstra, S. Legg, D. Hassabis, Human-level control through deep reinforcement learning. Nature **518**(02), 529–33 (2015)
27. S. Srinivasan, M. Lanctot, V.F. Zambaldi, J. Pérolat, K. Tuyls, R. Munos, M. Bowling, Actor-critic policy optimization in partially observable multiagent environments. CoRR, abs/1810.09026 (2018). http://arxiv.org/abs/1810.09026
28. T.D. Kulkarni, K. Narasimhan, A. Saeedi, J.B. Tenenbaum, Hierarchical deep reinforcement learning: integrating temporal abstraction and intrinsic motivation. CoRR, abs/1604.06057 (2016). http://arxiv.org/abs/1604.06057
29. P. Si, H. Liang, W. Wu, Y. Zhang, Joint resource management in cognitive radio and edge computing based industrial wireless networks, in *GLOBECOM 2017 - 2017 IEEE Global Communications Conference* (2017), pp. 1–6
30. F.S. Mohammadi, A. Kwasinski, QoE-driven integrated heterogeneous traffic resource allocation based on cooperative learning for 5G cognitive radio networks, in *2018 IEEE 5G World Forum (5GWF)*, Silicon Valley, CA, USA (2018), pp. 244–249. https://doi.org/10.1109/5GWF.2018.8516939
31. Z. Yang, K. Merrick, L. Jin, H.A. Abbass, Hierarchical deep reinforcement learning for continuous action control. IEEE Trans. Neural Netw. Learn. Syst. **29**(11), 5174–5184 (2018)

Chapter 11
The Role of Cognitive Radio Networks in Fifth-Generation Communication and Beyond

11.1 Next-Generation Wireless Communication Technologies

The newly-evolving wireless communication technologies, of which the cognitive radio networks (CRN) is a part, are generally classified as next-generation (xG) wireless communication technologies. The fifth-generation (5G) network is one prominent example of these new xG technologies. Beyond 5G, there are several other emerging xG technologies such as the internet-of-things (IoT) networks, the next-generation wireless sensor networks (xWSN), the device-to-device (D2D) communication networks (such as machine-2-machine communications and vehicle-2-vehicle communications) and many others. Most of these xG technologies are already reaching advance stages in their development and eventual deployment [1].

As the xG communication technologies develop and evolve, they face certain generic wireless communication challenges which they must strive to overcome. Some of these challenges are in the aspects of spectrum availability to drive the xG promises and possibilities [2], their ability to achieve or meet the high quality-of-service (QoS) requirements for most xG technologies [3], network reliability and robustness against network failures [4] and some others. There are ongoing works to address most of these challenges. Several telecommunication tools, such as the tools of optimisation, queueing theory and network restoration, are being investigated and employed to address the various daunting challenges of xG wireless communication technologies.

As more and more xG technologies evolve, an interesting aspect of these technologies is that they will influence one another in many significant ways. Certainly, the CRN, being one of the xG technologies already reaching advance stages in its development, will have significant impact and influence on other xG wireless communication networks. Some of the important roles that the CRN is playing and will continue to play as it interacts with other xG technologies are discussed in this chapter.

B. TJ Maharaj, B. S. Awoyemi, *Developments in Cognitive Radio Networks*, https://doi.org/10.1007/978-3-030-64653-0_11

11.2   The Role of Cognitive Radio Networks in Fifth-Generation Communication

The 5G communication is now being projected as the new standard of communication for emerging xG wireless communication networks. This standard of communication called 5G has important improvements over most other currently-applied standards of communication such as the third-generation (3G) standard, the long-term evolution (LTE) and LTE-Advanced standards, the WiMax standard, among others. The important improvements of 5G, and its many promises, have been shown to be significant enough to describe 5G as a generational shift in communication.

At the initial stages in the development of 5G, the definition and description on some of the minutest details of the 5G network were conflicting and sometimes compromising. However, as more and more research works on the subject are being carried out, and debates and deliberations are continuing, there has been a lot more progress in the harmonisation of thoughts, purpose, expectations and designs for the 5G networks. As advanced research works on 5G continues, experimental or trial deployments of 5G are now being concluded and full deployment of 5G is now taking place. Already, a large part of major cities in most technologically-advanced countries of the world currently use the 5G network.

The promises and expectations of 5G network are quite impressive. The 5G network is expected to achieve communication speeds of about 1 gigabyte per second and latency of less than 1 ms. Furthermore, 5G networks are expected to provide communication to cover almost all parts of the world and must have extremely high reliability. These promises and expectations are indeed worthy of making 5G to be considered and described as a generational shift in wireless communication [5]. Because of these massive promises of 5G, once it is fully operational, many 5G-based applications will emerge that will have very high social and economic value and far-reaching impact on all aspects of our lives [6]. More so, the 5G network will be one of the very important tools to help drive the realisation of the much-talked-about and highly anticipated hyper-connected world [7].

To help fulfil and realise its promises and expectations, without doubt, 5G networks will most likely require large bandwidths and, by inference, huge spectrum frequency bands for their operations. Unfortunately, as already mentioned over and over in almost all the Chapters of this book, the spectrum is a limited, non-ubiquitousness and highly-competitive communication resource and it is already currently scarce. The scarcity in spectrum availability is one of the potential limitations to the productivity of the 5G networks [8]. The CRN is being employed in 5G networks to assist in this regard.

The most significant role that the CRN plays in advancing the course of 5G networks is in relation to the spectrum. Very clearly, *by employing CRN in 5G, the limitations in spectrum availability for 5G applications can be significantly overcome* [9, 10]. More so, the CRN will help in *providing cognitive capabilities for the 5G networks* which will further help in achieving 5G goals, especially in the areas of latency, speed and network reliability.

11.3   The Role of Cognitive Radio Networks in Internet-of-Things Networking

The IoT has gained attention in recent times as the new paradigm of internet connectivity. As more and more xG communication technologies emerge, the IoT is also attracting and gaining equal attention and focus as some of these other xG communication and connectivity technologies [11]. In simple terms, the IoT is the new internet reality for the immediate and the near future. The IoT makes it possible to simultaneously and seamlessly interconnect many objects or 'things'. The 'things' that are interconnected in the IoT could be gadgets, machines, devices, buildings, structures, etc. These 'things' are connected through the internet to provide and achieve effective and efficient autonomous services, with little or no participation or intervention of human beings [12].

Fundamentally, the IoT is an offshoot of the internet [13]. To help differentiate the regular internet from the new IoT, the important difference is the fact that the term 'things' in the IoT has a wider and a more holistic meaning. In traditional internet considerations, the devices that are connected and used for the inter-networking are almost always computers. However, in the IoT, the 'things' do not necessarily have to be computers. The 'things' in the IoT are all matters with which it is possible to make connection, and/or such matters between which the exchange of information and communication can be carried out. Again, the 'things' in the IoT can be any uniquely identifiable fixed or mobile communicating object that is capable of collecting data, relaying information to other objects, collaboratively processing relayed information, and taking autonomous actions based on the information acquired or processed [8].

The interconnectivity that is achieved by the 'things' or objects in the IoT, alongside the inculcation of an high-speed software for timely data collection, processing and result analysis, empowers the IoT to render important and intelligent services to humans in all works of life. The services that are provided using the IoT are carried out in manners that are unachievable by the regular internet services and service providers [14]. Among others, the IoT provides top-of-the-range services in the areas of *communication, connectivity, providing direction, reporting, operations, providing warnings* and *carrying out timely interceptions*. These services are usually seamless, and require the barest amount of human participation [15].

As the IoT develops, it is becoming increasingly clear that it will depend on and work with several other xG technologies, particularly the xWSN, the 5G networks and the CRN. *The CRN will help in the development of smart devices for the IoT, in providing the much needed spectrum resource to drive the IoT operations and in optimising resource usage for the IoT. The cognitive capabilities provided through the CRN will be instrumental in driving autonomous, real-time and highly sensitive activities* of the IoT.

11.4   The Role of Cognitive Radio Networks in Advanced Wireless Sensor Networks

For many decades now, the use of sensing devices for a wide range of applications such as in health monitoring systems, in military operations, in home automations and in environmental monitoring systems is a well-established fact. In many of the use cases, the sensing devices are deployed with the capability of wireless connectivity. This is to enable an inter-networking and coordination of the activities of all the sensing devices located around the same geographical area. The operations of such interconnected sensing devices can therefore be expanded far and wide, even into remote locations. The wireless sensor networks (WSN), which is the inter-networking of wireless sensing devices, have been widely accepted because of the benefits of *flexibility reliability*, *scalability* and *ease of deployment* [16]. More so, the sensing devices in the WSN can easily exchange communication in a *tether-less and ad hoc manner* to help share their sensed data with neighbouring devices or nodes or to relay the results of sensing activities to a particular destination (or sink) device or node.

As the WSN evolve, advanced or next-generation wireless sensor networks (xWSN) are being developed to render new applications for addressing several new and complex communication challenges [17]. The new applications of the xWSN transcend the traditional use of the WSN. The commonest application of the regular WSN has been for sensing and tracking ambient conditions, especially the temperature, pressure, water and seismic levels of specific entities. Usually, the data collected by the sensing devices in the regular CRN are used for forecasting and/or monitoring some environmental conditions or natural disasters. The sensing devices in traditional WSN applications are also embedded into public structures such as buildings, bridges, roads and towers to help monitor them and to quickly discover any potential weaknesses or faults in time. While the WSN have been widely applied, the xWSN have even wider application and are now being deployed in almost all fields of human endeavour. For instance, in the field of health care, xWSN are now being used for monitoring the different vital signs of mobile patients, and are helping with early emergency responses for sudden, undetected or unplanned experiences.

The growing application of xWSN in virtually all areas of human endeavour will bring about a multiplication in the number of sensor nodes that will be placed in almost all locations to help collect data. It is most likely that the number of sensors that would be required globally for data collections would be very huge. Generally, the sensor devices or nodes used in the WSN and the xWSN do have certain limitations or constraints that may impair or impede their performance. Some of the limitations of most of the sensor nodes for the WSN and the xWSN are that they have limited energy capacities, low processing capabilities and short transmission ranges. These limitations of the sensors and sensor nodes in the WSN and the xWSN must be well addressed and overcome so as to help expand the functionality and operability of the WSN and the xWSN [18].

As new models of the xWSN are being developed and deployed, it is very important to employ improved sensors and sensor nodes that are energy-saving and highly proficient, and that have wide range of applications and functionality. The CRN is an important technology that can be incorporated into xWSN to help improve its operations. By incorporating the CRN into the xWSN *a number of the optimisation approaches and solutions from the resource allocation paradigms of the CRN would be useful in developing and deploying viable and very robust xWSN models.*

11.5   The Role of Cognitive Radio Networks in Smart Cities

A Smart City is formed by the combination of several smart environments, each managed by its own system, but with the ability to operate and/or communicate with other systems. One good example of such a system is smart homes. In a smart home, the various characteristics in the home are being monitored and relevant apparatus can be controlled remotely. If there are more than one smart home in an area, it is possible to network or interconnect the smart devices within a particular home with other smart home systems. This can then be used to achieve improved security and/or to form a smart neighbourhood watch system [19]. The inter-networking of the different systems can be achieved by employing xG communication technologies such as the 5G, IoT or CRN infrastructure.

It is quite exciting to know that Smart City systems are currently being established and advanced in different parts and environments of the world. Potential new smart system applications are being identified, alongside the infrastructure needs for supporting these future services [20, 21]. To make smart cities work, it is necessary to develop scalable device frameworks for the implementation of smart cities. These frameworks should be able to support a huge number of devices having high volumes of data to be communicated across many systems. Such communications within those frameworks must also be reliable and seamless. This is because, the ecosystem of smart cities can only be realised if there are systems of network infrastructure that can support the interconnection of billions (or maybe trillions or more!) of devices to provide seamless communication services. More so, these systems of network infrastructure must be good at transmitting data across the network at very great speeds with exceptionally low latencies to cater for some devices that require to execute real-time communications.

The major new and/or emerging technologies that will form the framework or base to support smart cities are the xG technologies such as the 5G networks, xWSN, IoT networks and certainly, the CRN. These xG technologies will play important roles in the design and implementation of smart cities, and in achieving an highly-interconnected world. The CRN, being a part of the xG technologies, will be instrumental in driving the realisation of smart cities. What is clear is that the new and quickly evolving aspects of smart homes, smart or e-security, smart or e-banking, smart or e-farming, smart or e-transport, e-health, etc. will all be involved in the realisation of smart cities. As these new aspects of modern innovative human-technology experiences are being designed and developed, they will rely heavily on the successful application and implementation of the CRN and many other xG wireless communication technologies to help drive their realisation.

11.6   The Role of Cognitive Radio Networks in 6G, 4IR and Other Emerging Technologies

Apart from the 5G networks, the IoT networks and the xWSN, there are several other xG technologies that are emerging, are being studied and are being deployed to drive near-future wireless communications. As it stands, there are now ongoing conversations on the *six-generation* of communication (6G). The 6G communication technology will operate in the terahertz (THz) band, and will be completely driven by high-level artificial intelligence operations. In 6G, there will be the proliferation of autonomous vehicles and carriers (such as the unmanned aerial vehicles), the use of optical wireless communication, quantum communication, wireless energy transfer and much more [22].

Furthermore, there is a continuous growth in industry 4.0 or the *fourth industrial revolution* (4IR). The 4IR is the ongoing automation-based manufacturing, industrial, infrastructure and/or production ideal for modern times [23]. Just as we have had various revolutionary leaps in the past in the aspect of industrialisation, the 4IR is the *digital revolution* leap of industrialisation for the current and immediate future. *Big data*, *cloud computing*, *artificial intelligence*, *robotics* and some others are interesting concepts being employed to drive 4IR.

Even further, *nanotechnology* applications are on the rise in medicine and health, agriculture, security and several other walks on human life. New technologies are being developed for banking, commerce, local and international trades, macroeconomics and so on. The developments in big data, cloud computing, artificial intelligence, robotics and others are being employed to drive new possibilities and to break new frontiers in the area of medicine and health, transportation, agriculture, housing, water supply, security, disaster management and much more.

In all of these, the CRN will continue to find relevance and application. This is because, most of these technologies and innovations will always depend on and employ the cognitive capabilities of the CRN, the spectrum and resource optimisation of the CRN and/or the simple and scalable structure of the CRN in achieving their goals. This makes the CRN to be a very useful technology in the pursuit and drive for modern and future wireless communication promises and possibilities.

11.7 Essentials for Practicable Application of Cognitive Radio Networks in Next-Generation Communications

As the CRN emerges, it must adapt for it to remain relevant and productive. If the CRN is to maintain its relevance, applicability and usefulness in helping to drive emerging xG communication technologies towards achieving their goals, there are some important points that must be put into consideration. The essential points to help maintain the usefulness and applicability of the CRN are discussed.

- **The CRN must be seen as a separate technology**. While there are indeed a host of other xG wireless communication technologies being advanced, the CRN must be seen as unique in its own way and different from any of the other emerging technologies. Of course, there are connecting links and well-defined common denominators, but yet, the CRN must not be confused or interchanged with any of the other emerging xG wireless networks.
- **None of the new and/or emerging technologies can work in isolation**. Thus, the CRN must be viewed from the perspective of being a part of a whole, and not as a solo or stand-alone technology for accomplishing all the xG communication demands. From this viewpoint, practical and experimental designs of the CRN are considered alongside other emerging technological designs.
- **There are already some ongoing developmental designs that incorporate the CRN with one or more of the other emerging technologies**. For instance, the CRN is being designed with the WSN in the new technology being referred to as the *cognitive radio sensor networks* (CRSN). In the hybrid technology of the CRSN, the advantages of the CRN is infused into the WSN to provide better overall sensor networking and operability. Several other hybrid designs of xG wireless communication networks are emerging and will continue to emerge. This is an highly exciting development and must be welcomed with open hands.
- **Not all the questions on the CRN or any of the other developing xG technologies have been fully or completely answered**. Hence, advances in and further research on the CRN and other xG technologies will most likely lead to newer results, better perspectives, improved workability and a more appealing overall outlook. This must be anticipated. As more and more questions are being asked, a lot more probing will be carried out and improvements in design, experimentation and ultimate implementation will be observed and eventually realised.

11.8   Summary of the Chapter

In summary, this Chapter has discussed the relevance and impact of the CRN to other emerging wireless communication technologies for the immediate and the near future. Truly, almost all the new and emerging technologies will be influenced in one way or another by the CRN. Since none of these technologies can work in isolation, a lot more work must be done to properly define how these technologies must be designed to collaborate, compensate, complement and consolidate one another for greater results and higher productivity.

References

1. A.S. Alfa, B. T. Maharaj, H.A. Ghazaleh, B. Awoyemi, *The Role of 5G and IoT in Smart Cities* (Springer International Publishing, Cham, 2018), pp. 31–54. https://doi.org/10.1007/978-3-319-97271-8-2
2. B.S. Awoyemi, B.T. Maharaj, A.S. Alfa, Resource allocation for heterogeneous cognitive radio networks, in *2015 IEEE Wireless Communications and Networking Conference (WCNC)* (2015), pp. 1759–1763
3. B.S. Awoyemi, B.T. Maharaj, A.S. Alfa, QoS provisioning in heterogeneous cognitive radio networks through dynamic resource allocation, in *Proceedings of the IEEE AFRICON* (2015), pp. 1–6
4. B.S. Awoyemi, A.S. Alfa, B.T. Maharaj, Network restoration for next-generation communication and computing networks. J. Comput. Netw. Commun. **2018**, 4134878, 1–13 (2018)
5. F.-L. Luo, C. Zhang, *5G Standard Development: Technology and Roadmap* (Wiley-IEEE Press, 2016), pp. 616–. http://ieeexplore.ieee.org/xpl/articleDetails.jsp?arnumber=7572796
6. A. Gohil, H. Modi, S.K. Patel, 5G technology of mobile communication: A survey, in *2013 International Conference on Intelligent Systems and Signal Processing (ISSP)* (2013), pp. 288–292
7. S. Kumar, G. Gupta, K.R. Singh, 5G: Revolution of future communication technology, in *2015 International Conference on Green Computing and Internet of Things (ICGCIoT)* (2015), pp. 143–147
8. B.S. Awoyemi, A.S. Alfa, B.T. Maharaj, Resource optimisation in 5G and Internet-of-Things networking. Wirel. Personal Commun. **111**(4), 2671–2702 (2020)
9. Z. Zhang, W. Zhang, S. Zeadally, Y. Wang, Y. Liu, Cognitive radio spectrum sensing framework based on multi-agent architecture for 5G networks. IEEE Wirel. Commun. **22**(6), 34–39 (2015)
10. X. Hong, J. Wang, C.X. Wang, J. Shi, Cognitive radio in 5G: a perspective on energy-spectral efficiency trade-off. IEEE Commun. Magaz. **52**(7), 46–53 (2014)
11. X. Xiaoli, Z. Yunbo, W. Guoxin, Design of intelligent internet of things for equipment maintenance, in *2011 International Conference on Intelligent Computation Technology and Automation (ICICTA)*, vol. 2 (2011), pp. 509–511
12. Z. Yu, W. Tie-ning, Research on the visualization of equipment support based on the technology of internet of things, in *2012 Second International Conference on Instrumentation, Measurement, Computer, Communication and Control (IMCCC)* (2012), pp. 1352–1357.
13. D. Singh, G. Tripathi, A. Jara, A survey of internet-of-things: Future vision, architecture, challenges and services, in *2014 IEEE World Forum on Internet of Things (WF-IoT)* (2014), pp. 287–292

14. P. Pereira, J. Eliasson, R. Kyusakov, J. Delsing, A. Raayatinezhad, M. Johansson, Enabling cloud connectivity for mobile internet of things applications, in *2013 IEEE 7th International Symposium on Service Oriented System Engineering (SOSE)* (2013), pp. 518–526

15. L. Zheng, S. Chen, S. Xiang, Y. Hu, Research of architecture and application of internet of things for smart grid, in *2012 International Conference on Computer Science Service System (CSSS)* (2012), pp. 938–941

16. I. Akyildiz, W. Su, Y. Sankarasubramaniam, E. Cayirci, Wireless sensor networks: a survey. Comput. Netw. **38**(4), 393–422 (2002). http://www.sciencedirect.com/science/article/pii/S1389128601003024

17. P. Rawat, K.D. Singh, H. Chaouchi, J.M. Bonnin, Wireless sensor networks: a survey on recent developments and potential synergies. J. Supercomput. **68**(1), 1–48 (2014). https://doi.org/10.1007/s11227-013-1021-9

18. R. Dou, G. Nan, Optimizing sensor network coverage and regional connectivity in industrial IoT systems. Syst. J. IEEE **PP**(99), 1–10 (2015)

19. X. Li, R. Lu, X. Liang, X. Shen, J. Chen, X. Lin, Smart community: an internet of things application. IEEE Commun. Mag. **49**(11), 68–75 (2011)

20. A. Zanella, N. Bui, A. Castellani, L. Vangelista, M. Zorzi, Internet of things for smart cities. IEEE Int. Things J. **1**(1), 22–32 (2014)

21. M.M. Rathore, A. Ahmad, A. Paul, S. Rho, Urban planning and building smart cities based on the internet of things using big data analytics. Comput. Netw. **101**, 63–80 (2016); Industrial Technologies and Applications for the Internet of Things. http://www.sciencedirect.com/science/article/pii/S1389128616000086

22. M.Z. Chowdhury, M. Shahjalal, S. Ahmed, Y.M. Jang, 6g wireless communication systems: Applications, requirements, technologies, challenges, and research directions. IEEE Open J. Commun. Soc. **1**, 957–975 (2020)

23. M. Peña-Cabrera, V. Lomas, G. Lefranc, Fourth industrial revolution and its impact on society, in *2019 IEEE CHILEAN Conference on Electrical, Electronics Engineering, Information and Communication Technologies (CHILECON)* (2019), pp. 1–6

Chapter 12
Future Opportunities for Cognitive Radio Networks

12.1 Problems Yet Unsolved in Cognitive Radio Networks

The most recent and/or ongoing developments in the cognitive radio networks (CRN), especially in the aspects of resource (spectrum and others) utilisation for improved productivity, have been discussed in this book. It is clear from the models and analyses of the CRN, as already discussed, that the CRN is indeed a very viable technology for achieving next-generation (xG) communication demands and goals [1, 2]. Therefore, ongoing research efforts and endeavours geared towards improving the CRN are by no means a waste of time/resources or an unproductive venture. Actually, a lot more work on the CRN still need to be carried out.

While more and more research works on the CRN are being carried out to help solve the various problems that have been identified and associated with the CRN (such as the problems with limited resources, interference, etc.), we note that there are yet a number of challenging problems still berating the CRN [3]. We note therefore that despite the best efforts of this book to identify and discuss all possible recent and/or ongoing efforts towards solving the problems of modern CRN, there are still a number of open-ended problems and challenges that require further probing and continuous investigations. Some of these problems are mentioned and discussed in this chapter.

In discussing the open-ended resource-related problems of the CRN, we identify specific problems that are associated with the use of the various tools and techniques being employed in solving the numerous RA problems for modern CRN (the tools and techniques have been discussed in the previous chapters of this book). Other generic problems that may limit the full realisation of the CRN are also identified and discussed.

12.2 Problems Associated with Optimisation in Cognitive Radio Networks

This book has discussed RA solutions through optimisation in a very comprehensive manner. Several key aspects of optimisation, as applicable to resource usage (spectrum and others), have been highlighted and new RA directions and results have been discussed. However, the topics and areas on RA optimisation for the CRN, as discussed in this book, are by no means exhaustive. There are still problematic aspects and research gaps on RA optimisation for the CRN that require further probing, attention and solutions [4]. We discuss some problematic aspects of the RA optimisation models and solutions for the CRN.

1. **Problem Development**
   One major challenge with the RA optimisation for the CRN is that, in a very broad sense, there seems to be a kind of disjointedness in RA problem development, and in the solution modelling and approaches being employed by the various researchers to address their RA problem developed for the CRN. In most cases, even when the RA problems to be investigated seem to be similar, the objective functions used by the various authors are usually different, the constraints may also differ, and the decision variables to be employed are almost certainly different as well. As a result, it becomes very difficult to find a form of coordination or focal point in the ideas being used to define and describe the RA problems for the CRN. Since the problem formulations are diverse dissimilar, it becomes very difficult to properly order the ideas put forth for investigating solutions for the RA problems in the CRN or to provide any particular standards for solving them.

2. **Problem Oversimplification**
   Another major challenge with the RA problem formulations and solutions in the CRN is the challenge of oversimplifying the problems. Many authors, in a bid to make the RA problems solvable, do neglect some important aspects or factors of the CRN. As a result of such oversimplification, most RA problems for the CRN are either unrealistic or impracticable in real-life scenarios. For instance, the aspect of heterogeneity in the CRN has been ignored by many early works in the CRN. More recent works on RA modelling and solutions for the CRN, such as the ones discussed in this books, now incorporate heterogeneity and other more practical CRN situations, to make the RA problems and solutions more useful. However, a lot more can still be done in this regard.

3. **Problem Generalisation**
   It is very true that there is still a big challenge with being able to capture, establish and explain all the concepts and details of the CRN in one single model. Because of the different architectural designs and broad classifications for the CRN (see Chap. 2), it is usually very difficult, if not impossible, to successfully develop RA models that can single-handedly accommodate all of the imports and aspects of the CRN. As it currently stands, authors develop smaller sizeable models

to address specific aspects of interest, while making reasonable and practical assumptions on other details of the CRN. This is like the most reasonable and viable approach to developing and analysing useful research models on the CRN. However, models that can fit a wide range of CRN assumptions, parameters, applications, performance details, etc., such as the models discussed in various chapters of this book, are greatly encouraged and must be pursued.

4. **Problem Standardisation**
   Another great challenge with RA problems and solutions in the CRN is the issue of standardisation. Currently, because the CRN is still very much a work-in-progress, there are no well-defined standards or standardisation procedures that are being used for the CRN. There are some good attempts, some still ongoing, to address this issue, though. For instance, the work in [5] is a summary of the work carried out by the IEEE 802.22 working group, specifically set up to describe a standard for wireless regional area networks (WRAN) that would make use of the TV white spaces (TVWS) in a manner that they do not interfere with other communication networks. While there are works still going on in this regard, the current reality is that the CRN do not yet have specific standards that have been fully established and well accepted with which they are to be designed and operated.

5. **Optimisation Complexity**
   The final RA optimisation-related problem of the CRN, as identified in this chapter, is that the field of optimisation itself, which is the main tool being employed to help solve RA-related problems in the CRN, is a highly diverse and very dynamic problem-solving tool with multiple dimensions of interpretation and application for obtaining solutions to problems. Because of its diversity and depth, arriving at a single, well-established, generalised or one-fits-all optimisation solution method or approach for addressing all RA problems in the CRN will always be a big challenge.

12.3   Problems Associated with Queueing Theory in Cognitive Radio Networks

The tool of queueing theory or systems (models and analysis) has been developed and employed for addressing several key aspects of RA in the CRN, as already discussed in a previous chapter of this book. However, despite the best efforts and works dedicated to developing appropriate queueing models for the CRN, there are numerous open problems that have been and/or are being identified for further discussions and considerations [6]. Some of these problems are discussed.

1. **Problem Complexity**
   One major problem with employing queueing systems to solving RA problems in the CRN is the realisation that most queueing models developed for the CRN are complex and difficult to analyse. Indeed, practical and/or realistic CRN models

are usually very complex because of the interactions and interdependencies of multifarious factors that combine to form the CRN. Often, these multifarious factors do result in complicated queueing models that are very challenging to analyse and possibly implement. More so, if these multifarious factors are considered concurrently, the resulting RA problems in the CRN may become too complex and possibly intractable in its analysis and implementation.

2. **Model Oversimplification**

   One of the greatest challenges with the application of queueing theory for solving RA problems in the CRN is the problem of oversimplification. Because of the huge complexities of the RA problems in the CRN, sometimes, in fact, many times, it is only reasonable to simplify such problems to have the best chance of getting them solved. However, it needs to be stated that the works that have contributed the most in helping to solve queueing-related RA problems in the CRN are those with the most generalised models and assumptions. The reason is that such generalised models, when compared to the simplified ones, are, to a higher degree, closer to real-life CRN situations and they do have a wider range of applications and usefulness. A good aspect of research works on queueing-related RA for the CRN, therefore, is to seek to extend and expand models in their scope and generality, in order to achieve better results for the CRN.

3. **Correlation between the States**

   In queueing analysis, most works have modelled the channel occupancy of primary users (PUs) by the ON–OFF process. A channel is ON when it is busy or occupied or unavailable for the secondary network. A channel is OFF when it is free or unoccupied or available for the secondary network. In such ON–OFF applications, there is usually no correlation between the ON and the OFF periods. However, for a more realistic modelling, the PU activities must be modelled such that there is some correlation between consecutive ON times and/or between the ON and the OFF times. It will be of great advantage if new investigations are geared towards the study of the effects of such alternative PU activity models on the queueing systems for solving RA-related problems in the CRN.

4. **Hybrid Modelling**

   Although the hybrid architecture of the CRN can be very challenging to analyse and implement, they provide greater capacity for the secondary network, as compared to either the underlay or overlay architectural designs of the CRN. However, there is a serious under-representation in the application of queueing models for the hybrid architecture in various studies on queueing models for RA in the CRN. The vast majority of the works have focused on applying queueing models to either the underlay or the overlay architectural representation of the CRN. The reason for this is that the queue modelling and analysis for the hybrid CRN representation is usually more complex than for the other CRN representations. However, the benefits such as the improvement in capacity gains for the hybrid architecture necessitates that more investigations on queueing models for the hybrid CRN be carried out. Some good examples of recent works in this regard are the works in [7, 8].

5. **Spectrum Hand-off**

   Queueing systems have been sparsely considered in the RA for the CRN, especially in the aspect of spectrum hand-off. The reason for this is that most works on the CRN do assume direct communication links, for example, between the SUs and the SUBS, or between a pair of SUs. As a result of such assumptions, singular queues (sometimes with possibly multiple servers corresponding to multiple channels) are usually employed in analysing the model. However, if the more correct assumption that the communication links have more than one intermediary (relay or cooperative) nodes between the source and destination secondary network devices (such as was considered and studied in [9]), the queueing system becomes more complex to analyse as each secondary network device would have its own queue. Thus, a more appropriate queueing model would be required to analyse the network. This is still an open topic for further research, as it may be particularly appropriate in developing ad hoc CRN systems.

12.4   Problems Associated with the Use of Stochastic Geometry in Cognitive Radio Networks

The adoption of the tool of stochastic geometry (SG) when modelling interference in wireless communications networks continues to attract a lot of attention due to its ability to characterise the spatial locations of users in more realistic terms. So far, it has been applied in many wireless networks such as the heterogeneous cellular networks, the CRN, the IoT networks, device-to-device networks, etc. However, there are some problems associated with the use of SG for interference management and control in the CRN. Some of these problems are discussed.

1. **Problems with the Poisson Point Processes**

   Although the Poisson point processes (PPPs) have received wide adoption among the point processes because of their tractability, careful observation of the CRN shows that independent PPPs may not properly capture the actual distributions of PUs and SUs. As a result, other point processes such as hardcore point processes are now being considered for the CRN. Nonetheless, SG is a very powerful tool capable of providing accurate interference models for any wireless network.

2. **Problems with Hardcore Point Processes**

   Hardcore point processes provide an opportunity to properly characterise interference in the CRN. However, the non-availability of the probability generating functional (PGFL) for hardcore point processes means their analyses are often approximated using the PGFL of PPP. With this, it is almost difficult to obtain tractable analyses when hardcore point processes are adopted without some assumptions and simplifications. This is one of the reasons why hardcore point processes have received less attention despite its suitability in the CRN.

3. **Problems with Interference Oversimplification**

   While proper interference management and control are central to an efficient and effective resource allocation process in the CRN, the analyses for several important performance metrics such as the signal-to-interference plus noise ratio (SINR), the probability of coverage, the probability of outage, the spectral efficiency, etc. are normally obtained based on the aggregate interference received at any test user within the considered network. In large-scale networks, the analysis for the probability distribution function (PDF) of aggregate interference is not known. Hence, the aggregate interference is generally modelled using the Laplace transform (LT) of its PDF or the equivalent characteristic function (CF) or the moment generating functions (MGF) [10]. However, the LT, CF and MGF are not sufficient to obtain exact performance metrics and the need to resort to various assumptions and simplifications have been adopted in the literature.

12.5 Problems Associated with the Use of Machine and Deep Learning in Cognitive Radio Networks

The introduction of the concepts of machine and deep learning into the CRN has shown promising signs in the aspects of resource management, improvement in cognitive capabilities, advancement in primary-secondary networks' coexistence and much more. However, there are still daunting challenges with the use of machine and deep learning in achieving more for the CRN. Some of the most generic problems associated with the application of machine and deep learning in the CRN are briefly discussed.

1. **Lack of Global Generalisation**

   One of the major challenges with the use of machine and deep learning, especially for spectrum and resource optimisation in the CRN, is the problem of lack of globally recognised and/or accepted models, methods and means for deep and machine learning. Being a new and active research area, there are still ongoing discussions on standardisation and generalisation for machine and deep learning. This currently affects its adaptation and application to modern CRN. However, we anticipate that as more clarity and consensus are demonstrated, the tool of machine and deep learning will find a lot more relevance as it will resonate more with modern CRN designs and applications.

2. **Anticipating Different Problem Cases**

   An important concern with the use of machine and deep learning in the CRN is how well they can anticipate and prepare for new and/or evolving problems. Being an emerging technology itself, the CRN is still a work-in-progress. Hence, deep and machine learning algorithms that will be most relevant to the CRN will be those that have the ability to imagine and anticipate different possible problem cases, prepare for them, solve them, learn from them and so on.

3. **Long-term Planning**

   Deep and machine learning models and algorithms that will be most appropriate for the CRN must have the ability to provide solutions that have both present and future near-future value and relevance. This can only be achieved through long-term planning and preparation. Therefore, deep and machine learning solutions for the CRN must be able to solve immediate problems, while anticipating future possibilities. In other words, they must plan long-term.

12.6   Other General Problems Still Associated with the Cognitive Radio Networks

There are some other general challenges with the CRN that may not be specifically related to any of the techniques or tools being employed for the CRN, but are still worth mentioning and discussing. We highlight some of those generic problems in this section.

1. **Problems associated with Network Interference**

   Till date, the problem of interference remains, arguably, the biggest challenge with the implementation and application of the CRN. While there are indeed ongoing research works that seek to properly characterise and address the problem of interference in the CRN and/or to mitigate its effects, such as the recent works in [4, 9, 11], a lot more work is still required to be done in this regard.

2. **Problems associated with Network Security**

   A serious problem with the CRN and most other emerging xG communication technologies is the problem of network security [12]. The employment of large number of devices in practical CRN designs will make the network very prone to the danger of being easily compromised. How to keep the CRN secure, despite possible malfunctions of the user equipment and/or malicious internal and external attacks on the network, is still an area that requires a lot more probing.

3. **Problems associated with the Cost of Implementation or Service Provisioning**

   As more and more prototypes of the CRN are being rolled out, the aspect of the economics of the technology has to be considered [12]. Stakeholders, telecommunication companies, investors, etc. would be interested in knowing the costs of implementing the CRN, the cost of providing services using the CRN, the marketability of the CRN, the return on investment for the CRN, etc. Providing adequate answers to these questions will indicate how viable the CRN is or will be.

4. **Problems associated with Network Policies**

   One of the problems associated with xG communication technologies is the problem of network policy [13]. The CRN will have to adjust to the diverse network policies being put in place and/or employed in various countries on the dynamic spectrum allocation and usage, maximum transmission powers, resource management and control, protocols and practices for network designs, network standards, etc. These network policies may aid or hinder the application and implementation of the CRN if they are not well handled.

5. **Problems associated with Network Failures**

   Another major challenge with RA in the CRN and most other emerging xG communication networks is the problem of network failures [14]. The CRN and other xG networks have promises of extremely high reliability, hence their application in very sensitive communication projects and prototypes, such as in vehicle-to-vehicle communications. For the CRN to be realisable, therefore, it must be built to be very robust against network failures. To achieve this, appropriate network restoration models and solutions must be investigated and employed in the CRN for practical, real-life applications.

6. **General Problems associated with the Implementation of New Network Technologies**

   All new telecommunication technologies usually have tethering implementation issues, and the CRN is no exemption [15]. Tethering problems are, in most cases, resolved as best practises and guiding principles are employed and followed in the implementation process of new technologies. The CRN will encounter some of these problems at the initial stages of implementation but will soon adjust and overcome those challenges.

12.7   Recommendations and Research Directions for Further Developments in Cognitive Radio Networks

The problems associated with the CRN, as identified and highlighted, can only be addressed through more targeted research works on the various aspects of the CRN. An important contribution of this book is to make useful recommendations that can help guide and strengthen further research works being carried out by different researchers and interest groups working on modern CRN designs and implementations. While some of the recommendations are quite generic for the CRN, a good number of the recommendations are geared towards improving further the overall performance of the CRN and making it more robust for applications beyond 5G. Thus, the recommendations are directed towards improving the results realised by the implementation of the various tools and techniques (such as optimisation, queueing theory, cooperative diversity, stochastic geometry and deep learning) being employed for modern CRN, as already discussed in this book.

12.7.1   Recommendations for Further Improvement in Optimisation and Queueing Modelling

The following recommendations are made to help improve the use of the tools of optimisation and queueing theory in their application to RA and other aspects of the CRN.

1. **Assumptions**

   While it is plausible that most works on the CRN will have to use simplifying assumptions in order to make their CRN problems tractable and/or manageable, it is recommended that such assumptions, especially for optimisation and queueing theory, must be generalised. Furthermore, there must be efforts to reduce such assumptions to the barest minimum for more comprehensive and fruitful CRN realisations.

2. **Standardisation**

   Generally, it will be of great advantage if well-defined standards are developed and employed for accurate and across-board CRN modelling. This is more important. It is therefore recommended that CRN standards be quickly agreed upon and established by all stakeholders for reasonable continuity in the process, and for tractability in the progress on the work on the CRN.

3. **Discrete-Time Queueing Models**

   In most of the queueing models that have been developed and employed to solve the RA problems in the CRN, the network characterisation has been mostly assumed to be in continuous-time. As a result, continuous-time Markov chains have been used to analyse and solve the RA problems. However, the more realistic and/or more practical consideration of the queueing-related RA problems in the CRN would be to assume that the CRN characterisation is in discrete-time. The implication of such assumption is that discrete-time Markov chains will be needed to characterise the RA models, and to analyse them. One good example of such characterisation and analysis of RA solutions using discrete-time Markov chains is found in the work in [7]. Representing the resulting queueing-based RA problems for the CRN in discrete-time will probably be more demanding but investigating and solving them as such, because they provide better results, is highly recommended.

12.7.2   Recommendations for Improving Cooperation-Based Solutions

Several new tools and techniques are being employed in modern CRN to address its limitations and improve overall productivity for the CRN. The concept of cooperative diversity and relaying to help mitigate the negative effects of stringent permissible interference temperatures in the RA solutions for the CRN was introduced and studied in a previous chapter. Usually, such cooperative models that are developed and employed for the CRN make the assumption that the users that are selected as cooperators or relays have no data of their own to transmit. While this is very possible, it may not be true at all times. Therefore, it should be possible to develop and analyse models that permit the cooperators that have selected to simultaneously transmit both their own data and the data of the other secondary network device that requires their help, in order to realise even better results for the CRN. Other tools and techniques that have been discussed have areas of improvements that can still be further investigated to help achieve more for the CRN.

12.7.3   Recommendations for Improving Interference Management Through the Use of Stochastic Geometry

The problem of interference has been shown to be one of the most crippling limitations to the realisation of the potentials of the CRN. To help manage and mitigate the activities of interference in the CRN, the use of the tool of SG has been established in a previous chapter, where several models were investigated and analysed. To further improve the use of the tool of SG in managing and mitigating interference in the CRN, a number of recommendations are provided.

1. **Need for more Stochastic Geometry-based Models**
   Despite the importance of the CRN, interference management and control in the CRN have received less attention compared to other wireless networks such as the heterogeneous cellular networks and Poisson wireless networks. The uniqueness of the CRN, however, means that the interference models developed for cellular networks may not be completely adopted in the CRN. For instance, the CRN allows SUs to access the channels belonging to PUs as long as their transmissions do not cause excessive interference in the primary network, while active SUs are expected to vacate the spectrum band before the arrival of any PU. This notion of priority does not exist in cellular networks where both users are licensed to use the channel [10]. Another important characteristic of the CRN is the dependent distributions among PUs and SUs. In order to avoid interference at the PUs, the technique of exclusion regions is important when modelling interference in the CRN. With the application of the tool of SG, the distribution of the SUs usually depends on the distribution of the PUs. Thus, the independent PPPs often assumed in cellular networks are not best suited for the CRN. Hence, more appropriate SG-based models that best capture the intricacies of the CRN are still being required.
2. **Spatio-temporal Analysis**
   In most models, a typical SU will continue to wait until an appropriate channel becomes available for its usage, while an interrupted SU can only resume or repeat its transmissions only when a channel becomes available. Similarly, PUs

may also experience retransmission as a result of the interference generated from nearby SUs. Hence, it becomes important to investigate the effect of transmission delay on users' experience in the CRN. A common assumption in the literature is that the buffers of all the primary transmitters (PTs) and the secondary transmitters (STs) are always full. Such an assumption fails to consider the relationships between the spatial and temporal distributions of users in the CRN [16]. An important future research area is in developing models and obtaining analytical results that capture both the spatial and temporal dependence among the PUs and the SUs when modelling interference in the network.

3. **Mobility among Users**

In many modern real-life systems, the majority of users are mobile and do expect uninterrupted wireless connection and service on their mobile devices. This means that the common adoption of stationary point processes may not be sufficient to capture the possible mobility in the network. However, mobility can complicate the analyses for various performance metrics. To date, mobility remains an open research area when characterising interference in wireless communication networks, and particularly in the CRN.

4. **Cognitive Channel Sensing**

Channel sensing is significant to interference management and control in any wireless network. As already established, channel availability is determined by SUs through channel sensing. However, SUs are expected to be low energy devices and may use most of their energy during the sensing phase, especially in the densely populated networks, with little energy remaining for its transmission phase. More appropriate channel sensing techniques that ensure the SUs use less energy during the sensing phase while reserving their energy for the transmission phase are useful and still very much needed.

5. **Capturing more practical Network Parameters**

Most of the SG-based interference models consider only a single PU channel in order to avoid complicated analysis. Also, power control—an important requirement of SUs when transmitting in the underlay CRN model—is largely neglected in the system model. The need to capture more realistic CRN parameters such as multiple PUs, multiple SUs, power control, etc. is important and still need to be further explored.

6. **Exact Analysis for Interference Parameters**

When the assumption of independent distribution is relaxed in the CRN, obtaining exact closed-form analysis for various metrics is difficult. Hence, obtaining tractable closed-form expressions without underestimating the intensity of users in the network is also another important area to consider. Another interesting area is the relaxation of the bipolar network model assumption. In such a case, the distance between any transmitter–receiver pair would no longer be a constant value, making it more practical. Such incorporations need to be further studied for the CRN.

7. **Probability Generation Functional for Hardcore Point Processes**

One interesting area that is still in need of further research is in obtaining the PGFL for hardcore point processes such as PHP, MHCP, etc. This will provide more understanding on hardcore point process applications in the CRN, and will enhance tractable analysis when these point processes are used for studying and mitigating interference in the CRN.

12.7.4 Recommendations for Improving Machine and Deep Learning Applications and Implementations

In the pursuit of more advanced AI-inspired strategies, applications and implementations for the CRN, it can be observed that there are potential problems that still require the proper attention and development of techniques of deep architecture for the CRN. Based on the comprehensive discussions on deep and machine learning for the CRN provided in a previous chapter, it is clear that there are key open challenges and future research directions for the application of deep and machine learning in the CRN. Some recommendations to help improve the development and use of deep architecture for the CRN are provided.

1. **More Research Works on Machine and Deep Learning**
   The chapter on deep learning has provided some AI-inspired learning architecture for exciting CRN applications. The discussions make us believe that advances in deep reinforcement learning that are yet to appear can help revolutionise the autonomous control of future mobile networks, improve mobile device energy consumption and to ensure network sustainability. The application and implementation of AI solutions for the CRN is therefore an exciting research area that require further probing. With the right research works, the most appropriate deep learning architecture for CRN applications will be designed, developed and implemented.
2. **Disruption-tolerant Networking**
   The fifth-generation (5G) of wireless communication and the IoT computer networking and connectivity are recent and highly promising developments in communication and computer networking [12]. With both technologies being key players towards achieving a smart and interconnected world, one important limitation towards achieving that promise, apart from the limitation of insufficient network resources, is the possibility of network disruptions. In order to address the challenge of network disruptions, disruption-tolerant networking solution models that would guarantee data delivery even when traffic is interrupted are required.

   Disruption-tolerant architecture do not need a continuous path to deliver data between end points and are a necessity for all kinds of terrestrial communications. Given that information/data transmission requires undisturbed media for it to propagate, performing this task is still categorised under model-deficit and algorithmic-deficit problems. Therefore, outlining how social network analysis and social properties can be exploited to design disruption-tolerant networking solutions is a promising direction.

To improve disruption tolerance in modern networks such as the CRN, the use of distributed collaboration among devices in the form of particle interaction and subjective logic can be applied and how each device becomes involved in collaboration with other devices can be evaluated. In this way, the device collaboration can then be converted into a clustering problem in which bipartite graphs can be used to obtain cliques and their evolution (that is, the change in the wireless network topology). Reinforcement learning strategies can then be applied to realise learning-based cognition techniques, subject to deficiencies such as the collection of sufficient data quality and statistical inference of network status.

It is important to note that, contrary to traditional wireless communications, in the CRN, disruption-tolerant networking is not that straightforward. Due to errors in spectrum sensing data, which cannot only cost SUs better resources but can also cause undue interference to PUs, the implementation of disruption-tolerant networking strategies are difficult in the CRN. In order to prevent wrong statistical inference, which can lead to wrong spectrum occupancy conclusions, it would be necessary to first derive alternate statistical and computational methods to minimise these errors. We particularly apply social network analysis and exploit social properties by merging individual spectrum sensing results with the dynamics that impact collaborative spectrum sensing. If spectrum sensing data is to be useful for decision-making in RA for the CRN, its time granularity must be of good quality. All of these are good research areas in the application of deep architecture for modern CRN implementation.

3. **Effective Communication at Federated Scale**

Vulnerabilities to network adversaries is still a pervasive problem in the CRN with malicious users targeting false alarm errors. Successful false alarms are of particular interest to malicious users since it creates spectrum holes that they can exploit [17]. Protecting the performance of the CRN against vulnerabilities to adversaries without sacrificing network performance is still an open problem in the CRN cryptography.

In terms of deep architecture for the CRN, the availability of generative adversarial networks (GAN) somehow eases this challenge of vulnerability to network adversaries. The GAN is a framework that trains generative models using the adversarial process. It consists of a generator and a discriminator and can produce data that follow a certain distribution [18]. The discriminator attempts to differentiate between real and fake data generated by the generator. The generator tries to generate plausible data in order to fool the discriminator into making mistakes, which introduces a min–max two-player game between the two. As a result of the min–max two-player game, the generator ends up generating data with the same distribution as the real data. This consequently makes deep learning algorithms vulnerable to adversarial attacks. Thus, constructing deep models that are robust to adversarial examples is imperative, but remains challenging.

4. **Multi-domain Knowledge Interpretation and Integration in Autonomous Applications**

As the CRN and its allied technologies, such as the 5G and the IoT, continue to steer xG operations and are becoming industry-ripe for AI-driven solutions, with their promise of lowering costs and boosting efficiencies through automation, there is a fundamental question that is still being ignored. The question is about the values/morals learning problem in machine morality, which is a problem that will be encountered in most autonomous applications, such as in *autonomous vehicles*. Beyond base connectivity and throughput, the definition of reward tends to include a lot of things such as the quality of service, quality of experience, security, reliability, etc. However, in terms of machine morality, the definition of the reward has to depart from this list by addressing the following questions:

- What is the objective function of future mobile and wireless networking?
- How should the reward that must be maximised in real-life networking problems be defined?

The intellectual field dedicated to these questions is called moral philosophy. In moral philosophy, the central questions are:

- What ought to be done?
- How should we live?
- Which actions are right and which ones are wrong?

The generic answer to these questions, quite frankly, is that it depends on the values. Therefore, as more and more advanced AI are being created, the status quo tends to depart from the realm of Atari games, where "reward" is cleanly defined in terms of points necessary to win a game and exist more in the real world. It should also depart from the currently explored classification problems to specify explicitly to a computer what a cat looks like. For example, autonomous vehicles have to make decisions with somewhat more complex definitions of the reward. Currently, the reward is still tied to getting safely to the destination, but if the vehicle is forced to decide between staying the course and hitting five pedestrians or swerving and hitting one, should it swerve? What if the one pedestrian is an innocent child, or a gunman on the loose? How does that change the decision, and why? Suddenly we have a much more complex problem when we try to define the objective function, and the answers are not as simple. Similarly, in the domain of machine morality, it might be difficult to specify exactly how to evaluate the rightness or wrongness of one action over another. However, it is possible for a machine to learn these values in some way, and this is called the *values learning problem*, and it may be one of the most important technical problems humans will need to solve before autonomous applications of deep learning in xG networks can be fully commissioned.

12.7.5 Other General Recommendations and Research Directions

A few other useful recommendations for further works on the CRN are provided.

1. **Introduction to 6G-based CRN**
   Beyond 5G, sixth-generation (6G) technology is poised to be the incoming generation of telecommunication. In 6G, communications will be on the terahertz frequency band, and will most likely be AI-based [19]. The CRN must begin to gear up towards finding relevance in 6G communication. As such, more research works that incorporate deep learning architecture and application in the CRN, and that explore other 6G-based concepts as they unfold, are highly recommended.
2. **Practical Application and Implementation of CRN to 4IR**
   The fourth industrial revolution (4IR) or industry 4.0 is taking centre stage as the newest revolution of industrialisation. Industry 4.0 is the digitalisation of the industrial/production/manufacturing space. The 4IR will thrive on automation, and will depend on advanced technological concepts of Big data, robotics, cloud computing, AI, among others [20]. Without doubt, the CRN will be an important technology to help drive 4IR, and some works that develop and describe their relationship and interactions are already available. However, more research works, especially on practical application and implementation of the CRN to 4IR, are still required.
3. **Commercialisation of CRN**
   As the CRN evolves, alongside other emerging technologies, one important aspect that will require continued research is how to commercialise and/or to improve the commercialisation of the CRN and its allied technologies. A technology is only as usable and productive as it is marketable and profitable. Investors and stakeholders will be willing to continue to contribute to the development of the CRN as longs as the return on investment makes it worth the while. Therefore, the marketability and profitability of CRN applications must be given its proper place in the research space for the CRN to maintain its relevance as a driving force for xG communications.

12.8 Concluding Remarks

This book has presented the CRN as one of the most promising paradigms of xG communication networks. The CRN is therefore attracting attention and gaining more and more recognition in the Telecommunication space. Its most amazing promise is that of alleviating the resource (particularly, the spectrum) scarcity problem, and most of the recent works, as discussed in the book, are geared towards developing new solutions for resource optimisation in the CRN. The CRN solutions discussed in this book detail new and improved ways by which the CRN may be able to optimally employ its scarce and limited resources to meet the demands of its numerous and diverse users, thereby helping the CRN achieve it much-acclaimed promises.

The ideas discussed in the book, as well as the findings presented, all come together to present a cogent, concise and well-coordinated response to some of the open-ended problems on the CRN, particularly with regard to optimal resource utilisation and improved productivity for the heterogeneous CRN. In conclusion, we project that, *if the tempo on recent/ongoing developments in the CRN is sustained, the CRN is set to take centre stage in the wireless communication space as the ideal prototype for achieving most of the emerging xG wireless communication and networking possibilities.*

References

1. B.S. Awoyemi, B.T. Maharaj, A.S. Alfa, Resource allocation for heterogeneous cognitive radio networks, in *Proceedings of the IEEE WCNC* (2015), pp. 1759–1763
2. B. Awoyemi, B. Maharaj, A. Alfa, Optimal resource allocation solutions for heterogeneous cognitive radio networks. Digital Commun. Netw. **3**(2), 129–139 (2017). http://www.sciencedirect.com/science/article/pii/S2352864816301043
3. B.S. Awoyemi, B.T.J. Maharaj, A.S. Alfa, Solving resource allocation problems in cognitive radio networks: a survey. EURASIP J. Wirel. Commun. Netw. **2016**(1), 176 (2016). https://doi.org/10.1186/s13638-016-0673-6
4. B.S. Awoyemi, B.T. Maharaj, Mitigating interference in the resource optimisation for heterogeneous cognitive radio networks, in *Proceedings of the IEEE 2nd Wireless Africa Conference (WAC)* (2019), pp. 1–6
5. C.-W. Pyo, X. Zhang, C. Song, M.-T. Zhou, H. Harada, A new standard activity in IEEE 802.22 wireless regional area networks: Enhancement for broadband services and monitoring applications in TV whitespace, in *Proceedings of the 15th International Symposium on WPMC* (2012), pp. 108–112
6. F. Palunčić, A.S. Alfa, B.T. Maharaj, H.M. Tsimba, Queueing models for cognitive radio networks: a survey. IEEE Access **6**, 50801–50823 (2018)
7. A.S. Alfa, H.A. Ghazaleh, B.T. Maharaj, A discrete time queueing model of cognitive radio networks with multi-modal overlay/underlay switching service levels, in *2018 14th International Wireless Communications Mobile Computing Conference (IWCMC)* (2018), pp. 1030–1035
8. A.S. Alfa, H. Abu Ghazaleh, B.T. Maharaj, Performance analysis of multi-modal overlay/underlay switching service levels in cognitive radio networks. IEEE Access **7**, 78442–78453 (2019)
9. B.S. Awoyemi, B.T. Maharaj, A.S. Alfa, Resource allocation in heterogeneous cooperative cognitive radio networks. Int. J. Commun. Syst. **30**(11), e3247 (2017). https://onlinelibrary.wiley.com/doi/abs/10.1002/dac.3247
10. H. ElSawy, E. Hossain, M. Haenggi, Stochastic geometry for modeling, analysis, and design of multi-tier and cognitive cellular wireless networks: a survey. IEEE Commun. Surveys Tutorials **15**(3), 996–1019 (2013)
11. S.D. Okegbile, B.T. Maharaj, A.S. Alfa, Interference characterization in underlay cognitive networks with intra-network and inter-network dependence. IEEE Trans. Mob. Comput. **PP**, 1–1 (2020)

12. B.S. Awoyemi, A.S. Alfa, B.T. Maharaj, Resource optimisation in 5G and Internet-of-Things networking. Wireless Pers. Commun. **111**(4), 2671–2702 (2020)
13. A.S. Alfa, B.T. Maharaj, H.A. Ghazaleh, B. Awoyemi, *The Role of 5G and IoT in Smart Cities* (Springer International Publishing, Cham, 2018), pp. 31–54. https://doi.org/10.1007/978-3-319-97271-8-2
14. B.S. Awoyemi, A.S. Alfa, B.T. Maharaj, Network restoration in wireless sensor networks for next-generation applications. IEEE Sensors J. **19**(18), 8352–8363 (2019)
15. A.S. Alfa, B.T. Maharaj, S. Lall, S. Pal, Mixed-integer programming based techniques for resource allocation in underlay cognitive radio networks: a survey. J. Commun. Netw. **18**(5), 744–761 (2016)
16. S.D. Okegbile, B.T. Maharaj, A.S. Alfa, Spatiotemporal characterization of users' experience in massive cognitive radio networks. IEEE Access **8**, 57114–57125 (2020)
17. A. Sivakumaran, A.S. Alfa, B.T. Maharaj, An empirical analysis of the effect of malicious users in decentralised cognitive radio networks, in *2019 IEEE 89th Vehicular Technology Conference (VTC2019-Spring)* (2019), pp. 1–5
18. I.J. Goodfellow, J. Pouget-Abadie, M. Mirza, B. Xu, D. Warde-Farley, S. Ozair, A. Courville, Y. Bengio, Generative adversarial nets, in *Proceedings of the 27th International Conference on Neural Information Processing Systems*. Ser. NIPS'14, vol. 2 (MIT Press, Cambridge, 2014), pp. 2672–2680
19. M.Z. Chowdhury, M. Shahjalal, S. Ahmed, Y.M. Jang, 6g wireless communication systems: Applications, requirements, technologies, challenges, and research directions. IEEE Open J. Commun. Soc. **1**, 957–975 (2020)
20. M. Peña-Cabrera, V. Lomas, G. Lefranc, Fourth industrial revolution and its impact on society, in *2019 IEEE CHILEAN Conference on Electrical, Electronics Engineering, Information and Communication Technologies (CHILECON)* (2019), pp. 1–6

.

Glossary

Affordability The quality of a product, service, device, etc. being able to be afforded because it is inexpensive and/or reasonably priced.

Capability It describes what can be done, achieved or accomplished, for example, through a technology or device.

Capacity The maximum achievable output of a technology or technological design, for instance, the cognitive radio network.

Data pre-processing A data mining technique that transforms raw data into an understandable or usable format.

Deep architectures Systems composed of multitude levels of non-linear operations, such as neural networks (NN) with many hidden layers.

Deep learning A subset of *machine learning* (ML). Deep learning is an artificial intelligence (AI) function with the goal of building systems that use intelligence to solve complex tasks.

Mobility The ability of a communication device to be able to communicate even while in motion.

Optimality The best or most effective result(s) obtainable, based on current or prevalent conditions (constraints) under which a technology such as the cognitive radio network operates. For example, in context, optimality of the cognitive radio network is achieved when the best performance (measured from the performance metrics such as the average data rates, throughput, outage probability, etc.) is realised, given the prevailing network conditions (that is, the available resources and the various constraints being considered).

Portability The quality of a component or device being handy and easy to carry about.

Productivity The measure of the efficiency or total output (yield) of a communication network or technology, such as the cognitive radio network.

Resourcefulness The ability of a technology such as the cognitive radio network, or a communication device, to find quick and smart ways to overcome its various limitations.

© The Author(s), under exclusive license to Springer Nature Switzerland AG 2022 237
B. TJ Maharaj, B. S. Awoyemi, *Developments in Cognitive Radio Networks*,
https://doi.org/10.1007/978-3-030-64653-0

Throughput   The total amount of data per unit time (total data rate) that is successfully transmitted by a communication network or technology, and is usually measured in bits per second (bps).

Ubiquity   The quality of something being available everywhere and in abundant supply.

Index

Printed in the United States
by Baker & Taylor Publisher Services